LAGRANGIAN FLUID DYNAMICS

The emergence of observing systems such as acoustically-tracked floats in the deep ocean, and surface drifters navigating by satellite, has seen renewed interest in Lagrangian fluid dynamics.

Starting from the foundations of elementary kinematics and assuming some familiarity of Eulerian fluid dynamics, this book reviews the classical and new exact solutions of the Lagrangian framework, and then addresses the general solvability of the resulting general equations of motion. A unified account of turbulent diffusion and dispersion is offered, with applications among others to plankton patchiness in the ocean.

Designed as a graduate-level text and work of reference, the book provides the first detailed and comprehensive analytical development of the Lagrangian formulation of fluid dynamics, of interest not only to applied mathematicians but also oceanographers, meteorologists, mechanical engineers, astrophysicists and indeed all investigators of the dynamics of fluids.

ANDREW BENNETT is Professor at the College of Oceanic & Atmospheric Sciences, Oregon State University.

OTHER TITLES IN THIS SERIES

All the titles listed below can be obtained from good booksellers or from Cambridge University Press. For a complete series listing visit

http://publishing.cambridge.org/stm/mathematics/cmma

Topographic Effects in Stratified Flows
PETER BAINES

Inverse Methods in Physical Oceanography
ANDREW F. BENNETT

Wave Interactions and Fluid Flows
A. D. D. CRAIK

Theory and Computation in Hydrodynamic Stability
W. CRIMINALE, T.L. JACKSON AND R.D. JOSLIN

Theory of Solidification
STEPHEN H. DAVIS

Dynamic Fracture Mechanics
L.B. FREUND

Finite Plastic Deformation of Crystalline Solids
K.S. HAVNER

Turbulence, Coherent Structures, Dynamical Systems and Symmetry
PHILIP HOLMES, J.L. LUMLEY AND GAL BERKOOZ

Acoustics of Fluid-Structure Interactions
M.S. HOWE

The Dynamics of Fluidized Particles
ROY JACKSON

Benard Cells and Taylor Vortices
E.L. KOSCHMIEDER

Ocean Acoustic Tomography
WALTER MUNK, PETER WORCESTER, AND CARL WUNSCH

Turbulent Combustion
NORBERT PETERS

Colloidal Dispersions
W.B. RUSSELL, D.A. SAVILLE, AND W.R. SCHOWALTER

Vortex Dynamics
P.G. SAFFMAN

The Structure of Turbulent Shear Flow
A.A. TOWNSEND

Buoyancy Effects in Fluids
J.S. TURNER

Reciprocity in Elastodynamics
J.D. ACHENBACH

Theory and Coputation of Hydrodynamic Stability
W.O. CRIMINALE, T.L. JACKSON, AND R.D. JOSLIN

Plasticity: A Treatise on Finite Deformation of Heterogeneous Inelastic Materials
SIA NEMAT-NASSER

LAGRANGIAN FLUID DYNAMICS

ANDREW BENNETT
College of Oceanic and Atmospheric Sciences
Oregon State University
Corvallis, USA

CAMBRIDGE
UNIVERSITY PRESS

CAMBRIDGE UNIVERSITY PRESS
Cambridge, New York, Melbourne, Madrid, Cape Town,
Singapore, São Paulo, Delhi, Mexico City

Cambridge University Press
The Edinburgh Building, Cambridge CB2 8RU, UK

Published in the United States of America by Cambridge University Press, New York

www.cambridge.org
Information on this title: www.cambridge.org/9780521853101

First published 2006
Reprinted 2007

A catalogue record for this publication is available from the British Library

ISBN 978-0-521-85310-1 Hardback

To John Joseph Mahony

Contents

Preface

Motivation

Leaves drifting in streams and blowing in the wind belong amongst our root impressions of the natural world. Plumes discharging into streams and pumping from smoke stacks symbolize our impact on that world. Thus it is baffling when as students we discover that fluid dynamics is seemingly exclusively investigated by measuring pressure at fixed points. The manometers in our first fluids laboratories plainly measure total stagnation pressure; the mechanical flow meters less obviously strike a dynamical balance between the torque of the partial stagnation pressure on the turbine blades and the torque of friction in the turbine bearings. Our hands and faces do feel the rush of a stream or the sweep of the wind, but these are brute sensations in comparison to the incisive information processing at work when our eyes follows a flow marker.

This is a book about the role of kinematics in fluid dynamics. The most revealing mathematical framework for developing kinematics is the Lagrangian formulation, long ago discarded for being unwieldy compared to the Eulerian formulation (Tokaty, 1971). Yet the discarded unwieldiness owes precisely to the richness of the kinematical information. This book might have been written any time in the twentieth century; the motivation now is the emergence of Lagrangian observing technology. The emergence is of course a reemergence; meteorologists have been routinely tracking weather balloons with theodolites since the nineteenth century. However, visual tracking and short transmitter life limit these data to being little more than local or Eulerian measurements of wind velocity and thermodynamic conditions. Radar, acoustics, satellite relays and satellite-based navigation changed all that in the late twentieth century. High-altitude balloons were tracked by satellites for days during the First GARP Global Experiment (WMO, 1977). Floats deep in the ocean are now tracked, effectively continuously, for months and even years with onboard hydrophones and moored arrays of pingers.

Aims

The Lagrangian formulation of fluid dynamics is not likely to replace the Eulerian formulation. Such is especially the case in computational fluid dynamics, although hybrid techniques are gaining ascendency. Rather, the Lagrangian formulation complements the Eulerian. Hence this book is not intended as a first course in fluid dynamics, and readers are assumed to have studied Eulerian fluid dynamics at least at the level of an introductory course having a scientific rather than engineering bent (Batchelor, 1973). Advanced calculus (Apostol, 1957) and Cartesian tensors (Jeffreys, 1931) are of necessity used extensively. The treatment of turbulence here assumes considerable preliminary familiarity with empirical, dimensional and statistical aspects (Lumley and Panofsky, 1964; Tennekes and Lumley, 1972). The purpose of the treatment developed here, indeed the purpose of the book as a whole, is to reveal the unifying power of the Lagrangian formulation for one of the great problems in physics. The further purpose is the drawing of a broader perspective for the analysis of the important new environmental data being collected with the emerging Lagrangian technologies.

Contents

The development of Lagrangian fluid dynamics falls naturally into major parts. Part I is concerned with the essence of the Lagrangian formulation, beginning with the kinematics of particles and the introduction of a sufficiently powerful notation for particle kinematics. The reader is advised against skipping lightly through this seemingly prosaic material. It quantifies the concept of conservation of particle identity, which is perhaps *the* intrinsically Lagrangian concept. The concept is captured by the *labeling theorem* of Kraichnan (1965), also known as *Lin's identity* (Lin, 1963). A striking corollary of this theorem is an exact expression for a generalized Lagrangian drift in a laminar flow and in each realization of a turbulent flow. Approximate drift formulae have long been the subject of speculation: here is the actuality. While the first candidate for a dependent variable in Lagrangian fluid dynamics is the particle path, the more readily observed structures are streamlines in a wind tunnel or towing tank, and streaklines downstream of sources such as discharge pipes and smoke stacks. In anticipation of the complexity of such flow, the introduction of statistical quantification is essential. The rudiments are found here; comprehensive treatments may be found elsewhere, for example Monin and Yaglom (1971, 1975). A few generalities may be made for single particle and particle pair statistics in homogeneous turbulence, that is, in turbulent flow which is statistically uniform in space. The problem of relating Eulerian and Lagrangian statistics is shown to be formally solved with functional or 'path' integrals.

The Lagrangian developments of dynamical principles into conservation laws for mass, momentum and energy should be familiar since they are found in most purportedly Eulerian texts. This familiarity underscores the greater directness and clarity of the Lagrangian formulation of Newtonion dynamics for fluids. The momentum equation in particular involves particle accelerations; these are second-order partial derivatives of particle position with respect to time elapsed since identification or release. Both Cauchy and Weber realized (Lamb, 1932) that one integration in time is immediately feasible. Pressure is supplanted by another scalar invariant, while Cauchy's vector invariant usurps vorticity. The Cauchy invariant reveals that neither the particle path nor the particle velocity is the intrinsic dependent variable in fluid dynamics; rather it is the Jacobi matrix or strain matrix of partial derivatives of position with respect to initial position, or with respect to whatever dependent variable identifies or labels the particle. Unlike particle position, the Jacobi matrix is invariant with respect to Galilean transformations of space. To split the hair, both the Eulerian equations of motion and the original Lagrangian equations of motion are Galilean invariant; it is their respective dependent variables of velocity and position which are not. Two Russian hydrodynamicists have recently pointed out that there is a matrix notation for the strain-based development (Yakubovich and Zenkovich, 2001). While this compact notation appears to offer no advantages for numerical computations, it has enabled its proponents to generalize the rotational wave of Gerstner (Lamb, 1932) to a new class of vortices.

Lagrangian fluid dynamics can be expressed as a variational principle; the invariance of the Lagrange density with respect to changes of particle labels leads to the fundamentally important conservation laws for Ertel's potential vorticity. The laws are derived here with the care that is owed, to the extent that the widely claimed naturality for the variational approach is not so compelling.

Lagrangian variables, both dependent and independent, need coordinates. All the coordinate options in the Eulerian formulation are available. The detailed forms for the Lagrangian equations in various coordinates suggest symmetries which are global in space: the familiar transformations representing rotations, which leave the equations invariant in form, are independent of position but may depend upon time. Of particular interest to meteorologists and oceanographers is the form of the Lagrangian equations in a uniformly rotating reference frame.

Real fluids are characterized by the constitutive relations between stress and strain. Newtonian fluids are defined by a linear relationship between the local stress tensor and the local rate of strain tensor. The locality is essentially Eulerian in nature. It is most simply expressed with Eulerian variables, and is particularly awkward in Lagrangian variables. Yet, again, the appearance of strain components, some of which may be rapidly growing, makes manifest the tendency for intensification of viscous stresses by differential particle motion. The locality of the Newtonian stress tensor is an expression of loss of memory, while the strength of the Lagrangian formulation is memory expressed as the retention of fluid particle identity. Which is the closer to reality: loss of memory or memory retention? While the Newtonian constitutive relation is of undisputed practical value, it is not so much a fundamental physical law as a "phenomenological" law, to use the language of Prigogine (1980). In other words, a fluid continuum is an abstraction, an unnatural artifice. Real air and real water consist of assemblies of molecules, obeying the fundamental laws of conservation of mass, momentum and energy. Viscous stresses are caused by nonequilibrium distributions of molecular velocity, as shown by the Chapman–Enskog deduction of the Navier–Stokes equations from Boltzmann's equation (which deduction must surely qualify as the "grand unified field theory" of the early twentieth century, and in fact of much of the world which really matters to us; see Chapman and Cowling, 1970). Alas, Boltzmann's *stosszahlansatz* is an admission of loss of memory at the molecular level (Thompson, 1988), so Lagrangian memory retention would seem to be in vain. Yet a complete topological description of the motion history of the macroscopic medium – the fluid continuum – demands a formulation in which memory retention is intrinsic. The crisis was created not by the development of the Lagrangian formulation, but initially by Boltzmann having randomizing Liouville's detailed microscopic description of molecular motion, and subsequently by Chapman and Enskog having taken moments of Boltzmann's distributional description.

Having declared that the fully Lagrangian formulation of fluid dynamics appears to offer no great numerical computational advantage, it would be desirable to be able to offer a great range of analytical Lagrangian solutions. Alas, there are only a few and these are also presented in Part II. Then again, there are about as few analytical Eulerian solutions, and strictly Eulerian numerical methods are being overtaken by semi-Lagrangian methods. It is curious that some problems admit explicit solutions in one formulation but not in the other. Irrotational flow past a circular cylinder admits an explicit Eulerian solution, but the Lagrangian solution is not explicit. The latter is presented here simply to make the point. On the other hand, there is no explicit Eulerian solution for the Gerstner wave. Nor are there for its generalizations, the Ptolemaic vortices of Yakubovich and Zenkovich (2001). There are analytical Lagrangian solutions for planar flows of real fluids, typically near flat plates. The Navier–Stokes equations for steady, incompressible viscous flow in a flat-plate boundary layer were simplified by Prandtl (Schlichting, 1960); as pointed out by Blasius (Schlichting, 1960), Prandtl's equations admit a single similarity variable, and the resulting nonlinear ordinary differential equation may be solved numerically. It is shown here that the Lagrangian form of Prandtl's equations admit two similarity variables, one of which includes time, leading to a pair of partial differential equations.

The Lagrangian formulation may be derived from the Eulerian by a transformation of variables, but the transformation is flow dependent. The two formulations are therefore sufficiently different from a mathematical point of view that the general solvability of the Lagrangian must be addressed. Indeed, the increasing interest in numerical Lagrangian fluid dynamics motivates the question: is the computer really computing a flow?

It has long been recognized that the Lagrangian formulation is natural for the analysis of conserved passive tracers. The formulation for diffusing tracers is greatly complicated by the appearance of the Jacobi matrix, but assuming the strain components are uniform in space permits an analytical solution. The assumption turns out to be a valid approximation for turbulent diffusion on a broad range of scales; the Lagragian solution developed throughout Part III provides a unifying theoretical development of the many subranges of homogeneous turbulence, and for diffusion of concentration gradients. The results go beyond mere dimensional consistency or similarity, correctly generating functional forms in subranges where alternative forms coexist, some of which are dimensionally consistent but wrong. Relative dispersion is shown to interact with spatially nonuniform plankton growth rates to destroy spatial patchiness in the plankton concentration. Part III, which offers a coherent and strictly Lagrangian presentation of turbulent diffusion ranging from microscales in liquid mercury to planetary scales in the stratosphere, is a completely reworked, reargued and augmented edition of an essay which first appeared in *Reviews of Geophysics* (Bennett, 1987).

No coherent presentation of Lagrangian fluid dynamics appears to have been offered prior to this book, but there are comprehensive accounts of a number of hybrid formulations. Their being both hybrid formulations and well described elsewhere, there is no need to cram them in here. The Abridged Lagrangian History Direct Interaction Approximation (Kraichnan, 1965; Frisch, 1995) is a perturbative development of stationary, isotropic turbulence. The formulation is indeed hybrid, having both Lagrangian and Eulerian aspects. "ALHDIA" yields the correct self-similar inertial subrange, while an analagous but strictly Eulerian formulation does not. ALHDIA also yields the viscous subrange observed definitively in a tidal channel in British Columbia (Grant, Stewart and Moilliet, 1962); as such the theory is one of the great unsung victories of middle twentieth century physics. The hybdrid Lagrangian formulation by Andrews and McIntyre (1978) permits Reynolds' averaging without loss of operator form. Applied to the atmosphere, the Lagrangian mean formulation facilitates the analysis of beams of internal waves. As repeatedly mentioned here, semi-Lagrangian numerical methods are pervading all of computational fluid dynamics; there are many introductory accounts (e.g., Durran, 1999). Finally come random flight models, which take the form of stochastic differential equations. The models are traditionally if mistakenly described as Lagrangian simulations. After all, they originated in Einstein's theory of the Brownian motion of minute but distinct particles in water (Pais, 1982). Stochastic differential equations are, as far as scientific content is concerned, no more than elegant algorithms for solving the associated diffusion equations, and the approximate closures that lead from the true probabilistic Lagrangian kinematics to the diffusion equations are profoundly suspect. Nevertheless, Rodean (1996) presents a comprehensive treatment of Monte Carlo simulation of turbulent diffusion. Only a very brief outline is included here, with application again to plankton dynamics.

The emerging Lagrangian observing technologies that so much motivate this book are reviewed in Part IV. The brief data survey includes many World Wide Web addresses for sites supporting these technologies, especially oceanographic surface 'drifters' and subsurface 'floats'.

It is shown in Chapter 7 of Part II that simple wave solutions of infinitesimal amplitude may be developed in the Lagrangian framework, just as in the Eulerian framework. Sums of the Eulerian wave solutions have routinely been fitted to real data, but on scales that deny the assumptions upon which the simple wave solutions are based such as an unbounded, uniform and constant medium of propagation. Yet we continue to torture the real atmosphere and real ocean on this Procrustean bed of simple wave expansions. The practice should be deemphasized in favor of inverse modeling, that is, finding fields that simultaneously give good fits to the finite amplitude equations of fluid dynamics in a realistically shaped and realistically stratified ocean basin, and to real ocean data. Both fits should be sought within hypothesized levels of error. By implication a dynamical model should include not just the equations of motion, initial conditions and boundary conditions, but also quantitative estimates of the errors in each of these component. Any failure to fit would most likely indicate an overoptimistically small prior for the dynamical errors, that is, something new would have been learned about ocean dynamics. The Eulerian theory of oceanic and atmospheric inverse modeling may be found elsewhere (Bennett, 1992, 2002; Wunsch, 1996). Emerging methods of Lagrangian inverse modeling and Lagrangian data assimilation in general are introduced in Part IV.

Again, dynamical investigations of fluid motion must move beyond approximate analytical solutions and "forward" numerical integrations of the dynamics, followed by simple comparisons with data. In the preferable inverse calculations, the dynamical constraints need not be satisfied exactly, so the conventional dynamical insights obtained by closely evaluating Eulerian term balances do not apply in general. In any event, the term balance approach is frought with difficulties on planetary scales, since many processes contribute to the balances as a rule and their respective roles in the balances vary substantially over the ocean basin in question. The search for local dynamical insights must be complemented with new and advanced skills at tracking fluid particles, estimating the convergence and divergence of these tracks and assessing the impact of such kinematic processes on the evolving pressure gradients, that is, on the dynamics. Much experience is needed, in order that insights may be drawn from the combination of inverse modeling and the Lagrangian perspective. For instance, Eulerian analysis of deep float data routinely involves unconstrained linear regression, for the estimation of the Eulerian pressure from the float tracks. Yet pressure is not the dynamically appropriate scalar field from the Lagrangian perspective. The ocean analyst should instead estimate the scalar field of Cauchy and Weber, in a manner consistent with Lagrangian kinematics and dynamics. If this book is effective, the next generation of physical oceanographers will be able to do so.

Ulterior motive

It should be evident that considerable amounts of mathematical needlepoint are required for the Lagrangian analysis of fluid dynamics. Today's students have a marvelous facility with computers, even though their manipulative skills are less honed. Equally admirable are the students of an earlier generation who could knock off the Tripos questions in Whittaker and Watson's

Modern Analysis. The author falls between the two generations, yet wishes to provide some opportunities and encouragement to today's students so that they might acquire some of the older masteries.

Corvallis, 2005.

Acknowledgments

Many figures have been reproduced from other sources with the kind permission of the authors and the copyright holders: Mark Abbott (Figures 15.8, 15.9), George Boer (Figure 11.17), Frank Champagne (Figure 12.1), Russ Davis (Figure 11.12), Joseph Er–El (Figures 11.10, 11,11), Howard Freeland (Figures 15.4, 15.5, 15.6, 15.7), Ann Gargett (Figures 12.2, 13.3), Christopher Garrett (Figure 15.1), Carl Gibson (Figure 13.2), Harold Grant (Figure 11.15), Robert Kraichnan (Figure 11.16), Joseph LaCasce (Figures 11.13, 16.1, 16.2), John Lumley (Figure 2.2), James McWilliams (Figure 11.7), John Middleton (Figures 15.2, 15.3), Robert Molinari (Figures 16.3, 16.4, 16.5), Pierre Morel (Figure 11.6) , James O'Brien (Figure 15.10), Allan Robinson (Figure 11.7), Paul Sullivan (Figure 11.9), American Association for the Advancement of Science (Figure 15.1), American Geophysical Union (Figures 10.1, 11.12, 12.3, 13.1, 15.2, 15.3, 16.3, 16.4, 16.5), American Institute of Physics (Figure 11.16), American Meteorological Society (Figures 11.6, 11.10, 11.11, 11.17, 12.1), Cambridge University Press (Figures 11.3, 11.9, 11.15, 12.2, 13.3), Elsevier (Figures 11.20, 15.8, 15.9, 15.10), Sears Foundation for Marine Research (Figures 11.13, 12.4, 15.4, 15.5, 15.6, 15.7, 16.1, 16.2), and Springer–Verlag (Figure 11.7). The quote from a WOCE Newsletter in Chapter 14 is made with thanks to Peter Niiler. All the reproductions and all the new line drawings were made by David Reinert. Boon Chua recomputed Figure 10.1.

William McMechan typeset the first draft of the manuscript. Drafting began when the author was supported by the Visiting Scientist Program of the University Corporation for Atmospheric Research, and was located at the Office of Naval Research Science Unit, Fleet Numerical Meteorology and Oceanography Center, Monterey CA. It is a great pleasure to acknowledge the hospitality and the productive environment provided by Captain Joseph Swaykos USN and his entire staff at "Fleet Numerical."

So much is gratefully owed to Ken Blake at Cambridge University Press for his cheerful patience, and for his sound advice. Many colleagues provided unwitting encouragement by remarking that a book on Lagrangian fluid dynamics was a good idea. One can only hope that the result is what they had in mind.

PART I

The Lagrangian Formulation

Introduction

Kinematics, statistics and dynamics: these are the basic elements of fluid dynamics. The Lagrangian formulation of the conservation laws for mass, momentum and energy are familiar to fluid dynamicists, as it is the natural way to extend Newtonian particle dynamics to fluids. Less familiar are: the conservation law for particle identity, which is effectively a definition of the independent Lagrangian variables; the path integral relationship between the statistics of random dependent Lagrangian variables and their Eulerian counterparts; the first integrals of Cauchy and Weber for the inviscid Lagrangian momentum equations, and the Cauchy vector invariant; the boundary conditions that must be imposed on compressible flow at boundaries defined by fluid particles (comoving boundaries), and the increasingly useful Lagrangian conservation law for momentum when the particle position is expressed in radial distance, longitude and latitude. The complexity of the divergence of the viscous stress tensor expressed in Lagrangian variables is undeniable, but the structure emphasizes the status of the Jacobi matrix as the Galilean invariant state variable that characterizes the flow. The Cauchy invariant is algebraically related to the Jacobi matrix and its Lagrangian time derivative; the conservation law for the Cauchy invariant in viscous flow is almost elegant.

1
Lagrangian kinematics

1.1 Conservation of particle identity

The essence of Lagrangian fluid dynamics is fluid particle identity acting as an independent variable. The identifier or *label* may be the particle position at some time, but could for example be a triple of the thermodyamic properties of the particle at some time. Time after labeling is the other independent variable. The fluid particle may not actually have been released into the flow at the time of labeling, but merely labeled with position or with some other properties at that time. Nevertheless, "time of release" will be used interchangeably with "labeling time." The subsequent position of the particle is a dependent variable, even though it may coincide with the independently chosen position of an Eulerian observer at the subsequent time. The Eulerian observer also employs time, after some convenient initial instant, as the other independent variable. Of course, a particle path can be calculated in the Eulerian framework by integrating velocity on the path, with respect to time. Indeed, the suppression or implicitness of this detailed path information is the basis of the relative simplicity of the Eulerian formulation. On the other hand, fluid velocity is readily calculated from the particle position in the Lagrangian framework by the local operation of particle differentiation with respect to time after labeling.

Conservation of particle identity is not an immediately compelling consideration in the Eulerian framework, but is fundamental in the Lagrangian. Bretherton (1970) correctly remarks that, since fluid particles having the same mass, momentum and energy can be interchanged without affecting the dynamics of the fluid, the particle identities are of no dynamical consequence. Yet kinematic information is the basis for the conceptualization of flow. Quantification of the kinematic principle of conservation of particle identity yields a striking identity which resembles but is entirely distinct from conservation laws for mechanical and thermodynamic properties. A first integral of the identity provides an exact formula for a generalized Stokes drift in laminar flow, and in each realization of a turbulent flow. The suitability of, for example, thermodynamic variables as particle identifiers does not require that they be conserved; it is their instantaneous values at the labeling time which are conserved for an individual particle.

The relationship between the Lagrangian and Eulerian formulations must be established with great pedantry, in order to establish the soundness of both. Consider, therefore, the fluid particle having the identifier or label a_i, $(i=1,2,3)$, such as its three-dimensional Cartesian coordinates, at some time s. At some later time t a Lagrangian observer, that is, an observer who moves with the particle, and who adopts a notation similar to that of Kraichnan (1965), records the position of the particle as $X_i(a_j, s|t)$. An Eulerian observer located at the position x_i at time t detects the particle if and only if

$$x_i = X_i(a_j, s|t). \tag{1.1}$$

See Figure 1.1.

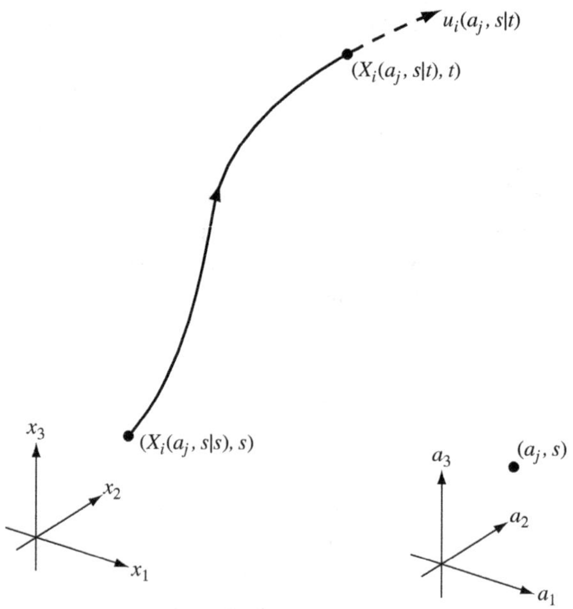

Figure 1.1 A fluid particle is given the label a_j at time s. Its position and velocity at time t are, respectively, $X_i(a_j, s|t)$ and $u_i(a_j, s|t)$. The label a_j is not necessarily the labeling position $X_i(a_j, s|s)$.

The Lagrangian velocity $u_i(a_j, s|t)$ is the particle velocity:

$$u_i(a_j, s|t) \equiv \frac{\partial}{\partial t} X_i(a_j, s|t). \tag{1.2}$$

Note that the partial derivative with respect to t is taken at fixed values for a_j and s, that is, the derivative is the Lagrangian partial in time. In the interest of notational simplicity, the same operator symbol $(\partial/\partial t)$ will be used subsequently for the Eulerian partial derivative in time, and the interpretation of the symbol will be made clear in the accompanying text. Subscripts will be used to distinguish thermodynamic partial derivatives of state variables, in the rare instances where such derivatives occur.

The labeling theorem Let q be any quantity associated with a fluid particle, such as density ρ, temperature T, or a velocity component u_i. The value of q at time t is denoted $q(a_j, s|t)$. *Assume that the label a_j is the particle position at time s.* Then, for any increment Δs in the labeling time s (see Figure 1.2),

$$q(X_i(a_j, s|s+\Delta s), s+\Delta s|t) = q(a_j, s|t), \tag{1.3}$$

since the labels are on the same path and they refer to the same particle. Expanding (1.3) and applying the definition (1.2) for the Lagrangian velocity yields (Kraichnan, 1965)

$$\frac{\partial}{\partial s} q(a_j, s|t) + u_k(a_j, s|s) \frac{\partial}{\partial a_k} q(a_j, s|t) = 0. \tag{1.4}$$

Note that there is an implied summation over the repeated index k in (1.4). The equation expresses that q is conserved along the characteristic direction

$$\frac{\partial a_k}{\partial s} = u_k(a_j, s|s)$$

in the (a_j, s) labeling space-time. This is the law of conservation of particle identity, or *labeling theorem.* ■

For example, choosing the quantity q to be any component u_i of the particle velocity,

$$\frac{\partial}{\partial s} u_i(a_j, s|t) + u_k(a_j, s|s) \frac{\partial}{\partial a_k} u_i(a_j, s|t) = 0, \tag{1.5}$$

and hence

$$u_i(a_j, t|t) = u_i(a_j, s|t) - \int_s^t u_k(a_j, r|r) \frac{\partial}{\partial a_k} u_i(a_j, r|t) dr. \tag{1.6}$$

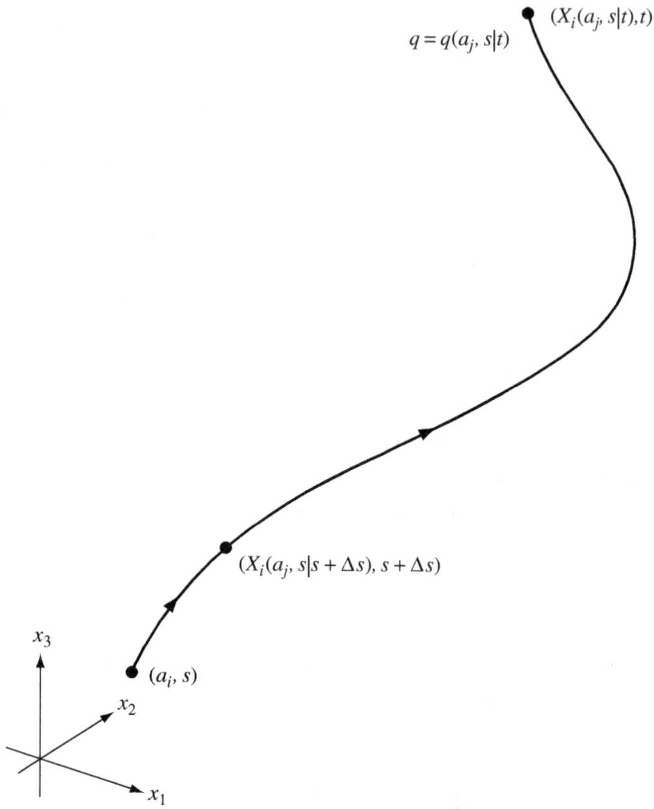

Figure 1.2 If a fluid particle is labeled by its position a_i at time s, then it could equally well be labeled by its position $a_i + u_i[a_j, s]\Delta s$ at time $s+\Delta s$. In particular, the value q for any state variable is the same for these two choices of labels.

When the label a_i is the particle position at the labeling time, as is the case here, it is convenient to introduce a special notation for the Lagrangian velocity at the labeling time:

$$u_i[a_j, r] \equiv u_i(a_j, r|r), \tag{1.7}$$

which is obviously the velocity recorded by an Eulerian observer at (a_j, r); this assertion will be carefully confirmed later. Introducing the Eulerian notation (1.7) into (1.6) yields

$$u_i(a_j, s|t) - u_i[a_j, t] = \int_s^t u_k[a_j, r] \frac{\partial}{\partial a_k} u_i(a_j, r|t) dr. \tag{1.8}$$

The relation (1.8) is an explicit expression for a *generalized* Stokes drift at $X_i(a_j, s|t)$ since, in general,

$$X_i(a_j, s|t) \neq a_i, \tag{1.9}$$

and thus the drift is the difference of Lagrangian and Eulerian velocities at different points on the one-particle path.

If the Eulerian velocity is solenoidal:

$$\frac{\partial}{\partial x_k} u_k[x_j, t] = 0, \tag{1.10}$$

then the drift is the spatial gradient of a mixed Eulerian–Lagrangian "prediffusivity:"

$$u_i(a_j, s|t) - u_i[a_j, t] = \frac{\partial}{\partial a_k} K_{ik}(a_j, s|t), \tag{1.11}$$

where

$$K_{ik}(a_j, s|t) = \int_s^t u_k[a_j, r] u_i(a_j, r|t) dr. \tag{1.12}$$

Notes

(i) The above formulae hold for a laminar flow, and for individual realizations of a turbulent flow; in particular the "prediffusivity" K_{ik} has not been averaged over an ensemble.
(ii) The product in the integrand involves total velocities, rather than departures from ensemble means.
(iii) The prediffusivity is asymmetric: $K_{ik} \neq K_{ki}$.
(iv) Equation (1.12) is hardly surprising: if the velocities in the integrand are known, then so is the drift (1.11). Nevertheless, it is instructive to assess the data needed to evaluate K_{ij}: a current meter (to use oceanographic terminology) must be deployed at a_i for $s < r < t$, and floats must be released at a_i at each time r in that interval: see Figure 1.3.

Exercise 1.1 Consider labeling by the particle position at the labeling time. Show that for any particle property q,

$$q(a_i, s|t) = q\big[X_i(a_j, s|t), t\big]. \tag{1.13}$$

Hint: let $q\big[X_i(a_j, s|t), t\big] \equiv q\big(X_i(a_j, s|t), t|t\big) = Q(a_i, s|t)$, say. Verify that $Q(a_i, s|t)$, like $q(a_i, s|t)$, satisfies the labeling theorem (1.4), and note that $Q(a_j, t|t) = q(a_i, t|t)$. This exercise establishes that the Lagrangian value of q at time t is the Eulerian value at the particle position at that time. Thus $q[x_i, t]$ is aptly named the Eulerian value. □

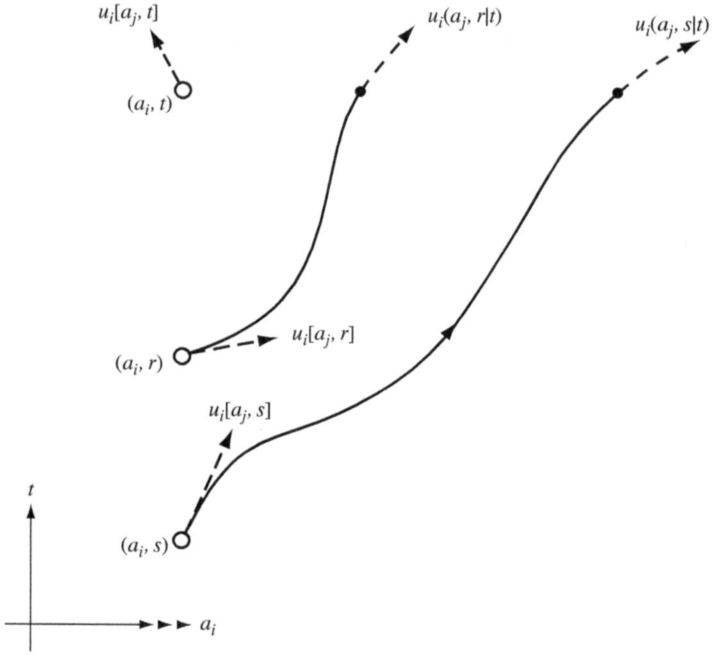

Figure 1.3 Evaluation of the generalized drift (1.11) requires that a current meter be deployed at position a_i for $s \leq r \leq t$, and that labeled fluid particles be released at a_i throughout the same time interval.

Exercise 1.2 (Lin, 1963) The notation of the labeling theorem, like that the path function $X_i(a_j, s|t)$, can be reversed for further illumination. Let a_i be the label, at time s, of a particle observed at position x_j at time t; that is, $a_i = A_i(x_j, t|s)$. Show that the "total" or "material" derivative of the labeling function A_i vanishes identically:

$$\frac{\partial}{\partial t} A_i(x_j, t|s) + u_k[x_j, t] \frac{\partial}{\partial x_k} A_i(x_j, t|s) = 0. \tag{1.14}$$

Note that, unlike Kraichnan's equation (1.4), Lin's equation (1.14) holds not only for labeling by position at time s, but for arbitrary labeling at that time. □

Exercise 1.3 Extend the labeling theorem to labels other than the particle position at the labeling time, according to the following principle: for a fluid particle at position x_i at time t, the value of any particle property q is independent of the time s at which the particle is assigned the arbitrary label a_j. Verify that the original theorem (1.4) does obtain when the label is in fact the particle position at the time of release. Alternatively, express any label as a function of the release position and invoke the original labeling theorem. Reconcile these extensions. Finally, given Lagrangian kinematics labeled by a_j at time s, relabel by b_j at time r. □

Exercise 1.4 Consider a Lagrangian flow formulation having arbitrary labels a_j, that is, labels other than the particle position $X_j(a_k, s|t)$ at the release time $t=s$. Express the Eulerian velocity in terms of the Lagrangian kinematics. Establish the aptness of the construction of Eulerian fields from Lagrangian fields having arbitrary labels. □

Exercise 1.5 Assume that a particle path of the form $X_i = X_i(a_j|t)$ is known to be a solution of the Lagrangian equations of fluid dynamics, for some label a_i. Is $X_i = X_i(a_j|t-s)$ also a solution, for some time s? Show that the labeling theorem may be used to extend the known solution to a family of solutions of the form $X_i = X_i(a_j, s|t)$. □

1.2 Streaklines, streamlines and steady flow

Fluid flow tends to be time dependent, and is most naturally made visible with streaklines. These are neither particle paths nor streamlines, except for steady flow in which all three are identical.

Exercise 1.6 A streakline is the locus, at one time t, of fluid particles released at the position x_i at previous times r in some interval $s \le r \le t$. Express streaklines with Lagrangian notation. A streamline is a path everywhere tangential to the local fluid velocity, at one time t. Express streamlines with Lagrangian notation. Illustrate planar particle paths, streakline and streamlines with a single perspective sketch in the (x_1, x_2, t) space-time. □

Flow is defined to be "steady" if Lagrangian values are invariant under time translation:

$$q(a_i, s|t) = q(a_i, s-T|t-T), \tag{1.15}$$

for some time shift T. The left-hand side of (1.15) can depend on s and t only in the combination $t-s$. We may then define

$$q(a_i|t-s) \equiv q(a_i, s|t). \tag{1.16}$$

The "streamline" $X_i(a_j|t-s)$ is the sole particle path through $X_i(a_j, s|s)$:

$$X_i(a_j|t-s) = X_i(a_j, s|t). \tag{1.17}$$

Exercise 1.7 Assuming that particles are labeled by their positions a_j at time s, show that on a streamline in steady flow,

$$u_i(a_j|t-s) = \left(\frac{\partial}{\partial a_k} X_i(a_j|t-s)\right) u_k(a_j|0). \tag{1.18}$$

That is, the velocity on the streamline is the "strained initial value". Hint: use the labeling theorem. Is (1.18) a linear relationship? □

In general, the matrix of "Lagrangian strains"

$$J_{ij}(a_k, s|t) \equiv \frac{\partial}{\partial a_j} X_i(a_k, s|t) \tag{1.19}$$

is the Jacobi matrix for the transformation $a_j \to X_i$. The Lagrangian formulation is useful only so long as the determinant of this transformation, or Jacobi determinant, does not vanish.

Recall that for labeling by release position, the Eulerian velocity is

$$u_i[x_j, t] \equiv u_i(x_j, t|t). \tag{1.20}$$

If the flow is steady, then

$$u_i(x_j, t|t) = u_i(x_j|0), \tag{1.21}$$

and the Eulerian velocity is independent of time:

$$u_i[x_j, t] = u_i[x_j]; \tag{1.22}$$

thus it suffices to find the Eulerian velocity at time $t = s$. The Eulerian and Lagrangian velocities coincide at that time:

$$u_i[x_j] \equiv u_i(x_j|0). \tag{1.23}$$

Exercise 1.8 Show that in steady flow, particle paths are also streaklines and streamlines. □

Now consider an ideally conserved quantity such as entropy η. That is,

$$\frac{\partial \eta}{\partial t} = 0. \tag{1.24}$$

If the flow is steady: $\eta(a_j, s|t) = \eta(a_j|t-s)$, then by the labeling theorem

$$u_k[a_j]\frac{\partial}{\partial a_k}\eta(a_j|t-s) = 0. \tag{1.25}$$

This startling conclusion may be reconciled to the Eulerian expression of steady convection:

$$\begin{aligned} u_k[a_j]\frac{\partial}{\partial a_k}\eta(a_j|t-s) &= u_k[a_j]\frac{\partial}{\partial x_m}\eta[X_i]\frac{\partial}{\partial a_k}X_m(a_j|t-s) \\ &= -\frac{\partial}{\partial s}X_m(a_j|t-s)\frac{\partial}{\partial x_m}\eta[X_i] \\ &= \frac{\partial}{\partial t}X_m(a_j|t-s)\frac{\partial}{\partial x_m}\eta[X_i] \\ &= u_m[X_i]\frac{\partial}{\partial x_m}\eta[X_i] = 0. \end{aligned} \tag{1.26}$$

Note:

(i) Relations (1.14) and (1.23) have been applied to η, and the labeling theorem has been applied to the steady particle path $X_m(a_j|t-s)$.

(ii) There is seemingly more information in (1.25) than in the rightmost equality of (1.26), since the former refers to the Lagrangian gradient of η at times other than the labeling time. However, since η is conserved and since steady flow is assumed, $\eta = \eta(a_j|t-s) = \eta(a_j|0) = \eta[a_j]$.

1.3 Local kinematics

The definition (1.2) for the Lagrangian velocity $u_i(a_j, s|t)$, the definition (1.7) for Eulerian velocity $u_i[x_k, t]$, and the identity (1.13) lead to the well-known relation

$$\frac{\partial}{\partial t}X_i(a_j, s|t) = u_i[X_k(a_j, s|t), t]. \tag{1.27}$$

Assuming that $X_i(a_j, s|s) = a_i$, that is, the particle is labeled by its position at time s, assuming smoothness of the Eulerian velocity field, and expanding in a Taylor series about the local reference point a_i^* for small $t-s$ yields

$$\begin{aligned} \frac{\partial}{\partial t}X_i(a_j, s|t) =& u_i[a_j^*, s] + \left.\frac{\partial}{\partial x_k}u_i[x_j, s]\right|_{x_j=a_j^*}\left(X_k(a_j, s|t) - a_k^*\right) \\ &+ O(|a_j - a_j^*|^2) + O(t-s). \end{aligned} \tag{1.28}$$

The Eulerian rate of strain tensor at $[a_i^*, s]$ may be decomposed into symmetric and skewsymmetric tensors:

$$\left(\frac{\partial u_i}{\partial x_k}\right)^* = \frac{1}{2}\left(\frac{\partial u_i}{\partial x_k} + \frac{\partial u_k}{\partial x_i}\right)^* + \frac{1}{2}\left(\frac{\partial u_i}{\partial x_k} - \frac{\partial u_k}{\partial x_i}\right)^*. \tag{1.29}$$

The skew tensor may be expressed in terms of a vector product:

$$\left(\frac{\partial u_i}{\partial x_k} - \frac{\partial u_k}{\partial x_i}\right)^* = -\epsilon_{ikl}\omega_l^*, \tag{1.30}$$

where the alternating tensor $\epsilon_{ikl} = 1$ for $i = 1, k = 2, l = 3$, etc. (Jeffreys, 1931), and ω_l^* is the value at $[a_i^*, s]$ of the Eulerian vorticity ω_l:

$$\omega_l = \epsilon_{lmn}\frac{\partial u_n}{\partial x_m}. \tag{1.31}$$

Transforming to a new spatial variable ξ_i, according to

$$X_i = a_i^* + (t - s)\left(u_i^* - \frac{1}{2}\epsilon_{ikl}\omega_l^*\xi_k\right) + \xi_i, \tag{1.32}$$

the local relation 1.28 becomes

$$\frac{\partial \xi_i}{\partial t} = e_{ik}^*\xi_k + O(|a_j - a_j^*|^2) + O(t - s) \tag{1.33}$$

where e_{ik}^* is the local value of the symmetric rate of strain tensor

$$e_{ik} = \frac{1}{2}\left(\frac{\partial u_i}{\partial x_k} + \frac{\partial u_k}{\partial x_i}\right). \tag{1.34}$$

Note that $\xi_i = a_i - a_i^*$ at $t = s$. Again, all the Eulerian fields are evaluated at the reference point a_i^* and at time s. The transformation (1.32) consists of an infinitesimal translation with the local velocity u_i^*, plus an infinitesimal rotation with the local angular velocity $\omega_l^*/2$. A further transformation to the principal axes of the symmetric tensor e_{ik}^*, and (1.33) is diagonalized:

$$\frac{\partial \xi_i{}'}{\partial t} = \lambda_i\xi_i' + \dots, \tag{1.35}$$

where ξ_i' is the component of displacement in the ith principal direction, and λ_i is the ith principal moment. The trace of a matrix is an invariant, thus

$$\lambda_1 + \lambda_2 + \lambda_3 = e_{kk} = \frac{\partial u_k}{\partial x_k}. \tag{1.36}$$

It follows that for a solenoidal Eulerian velocity field or "incompressible flow," the sum of the eigenvalues λ_i vanishes. If all vanish, then the flow is stagnant in the transformed coordinates, and the particle motions in the original coordinates are circles superimposed on a uniform translation. Assume to the contrary that at least one eigenvalue is positive and one is negative. The corresponding principal axes are, respectively, a *dilatation* axis and a *compression* axis passing through the reference point a_i^*. According to (1.33), other particles released near a_i^* approach the axis of greatest dilatation, asymptotically for large $t-s$. Note that, consistent with the preceding approximations, the Taylor series expansion of the Jacobi matrix for small elapsed time $t-s$ is

$$J_{ij}(a_k, s|t) = \delta_{ij} + (t-s)\Big(e_{ij}[a_k, s] - \frac{1}{2}\epsilon_{ijl}\omega_l[a_k, s]\Big) + O(t-s)^2, \qquad (1.37)$$

to first order. Thus the time evolution to this order of accuracy is determined by the Eulerian symmetric rate of strain tensor and the Eulerian vorticity, both evaluated at the labeling position and time. Higher order terms are determined by the pressure field, that is, by the dynamics of the fluid.

The preceding local analysis of particle kinematics is traditional, but is used to great effect in the study of turbulent diffusion by Batchelor (1959), and is the basis for much topological investigation (e.g., Ottino, 1989). The analysis is essentially Eulerian, as the characteristics of the particle motion are all determined by the spatial gradients of the Eulerian velocity at the original labeling position.

2

Lagrangian statistics

2.1 Single-particle, single-time statistics

Velocity provides the fundamental statistics in the Lagrangian formulation of fluid dynamics. Displacement is an integral of velocity, and so necessarily has nonstationary statistics even if those of velocity are stationary. Consider therefore the statistics of the Lagrangian velocity $u_j(a_k, s|t)$ at time t, for a particle having the label a_k at time s. Ignoring questions of convergence, assume that the Lagrangian velocity has a Taylor series expansion about s in powers of $t-s$:

$$u_j(a_k, s|t) = \sum_{n=0}^{\infty} \frac{1}{n!} \frac{\partial^n}{\partial t^n} u_j(a_k, s|t)\bigg|_{t=s} (t-s)^n \,. \tag{2.1}$$

For any labeling, the Lagrangian and Eulerian velocities are related by $u_j(a_k, s|t) = u_j[X_i(a_k, s|t), t]$. Hence

$$\frac{\partial^n}{\partial t^n} u_j(a_k, s|t)\bigg|_{t=s} = \left(\frac{\partial}{\partial t} + u_l[\xi_i, t]\frac{\partial}{\partial \xi_l}\right)^n u_j[\xi_i, t]\bigg|_{(\xi_i = X_i(a_k, s|s),\, t=s)} \tag{2.2}$$

for all $n \geq 0$. Assume that the Eulerian velocities at multiple points are statistically homogeneous, or in other words assume that the expectation value of any function of multiple $u_j[x_i, t]$ is independent of the absolute positions x_i. Then by (2.1) and (2.2) the Lagrangian velocity with any single label is also statistically homogeneous. That is, the expectation value of any function of $u_j(a_k, s|t)$ is independent of the label a_k (Lumley, 1962).

Let $\mathcal{U}$ be any function of u_j, and integrate the composed function $\mathcal{U}(u_j(a_k, s|t))$ over a parcel of fluid occupying the *spatial* domain V^s at the labeling time s:

$$I = \int_{V^s} \mathcal{U}(u_j(a_k, s|t))dV^s \,. \tag{2.3}$$

The volume element dV^s is $|J_s^s|dW$, where $dW = da_1 da_2 da_3$. The factor J_s^s is the value, at time $t = s$, of the Jacobi determinant J_s^t for the labeling transformation $a_j \to x_i = X_i(a_j, s|t)$. That is, $J_s^t = \det(\partial X_i / \partial a_j)$. Note that a_j need not be the particle position at time s, thus the element of measure dW for the labels is not necessarily a volume element. The Lagrangian velocity field in (2.3) may be expressed in terms of the Eulerian velocity field $u_j[x_i, t]$ and the particle paths $X_i(a_k, s|t)$:

$$I = \int_{V^s} \mathcal{U}(u_j[X_i(a_k, s|t), t])dV^s . \tag{2.4}$$

Changing variables from the label a_k to the position $x_i = X_i(a_k, s|t)$ yields

$$I = \int_{V^t} \mathcal{U}(u_j[x_i, t]) \frac{|J_s^s|}{|J_s^t|} dV^t , \tag{2.5}$$

where V^t is the spatial domain, at time t, of the parcel of fluid which occupies the spatial domain V^s at time s: see Figure 2.1. The volume element dV^t is $dx_1 dx_2 dx_3$, which is related to dV^s by

$$dV^t = |J_s^t| da_1 da_2 da_3 = \frac{|J_s^t|}{|J_s^s|} dV^s . \tag{2.6}$$

It will be assumed that the Jacobi determinant remains positive for all finite time.

Exercise 2.1 Show that

$$\frac{\partial}{\partial t} J_s^t(a_k, s|t) = \left(\frac{\partial}{\partial x_l} u_l[x_i, t] \right) J_s^t(a_k, s|t) . \tag{2.7}$$

Hint:

$$\frac{\partial(u_1, X_2, X_3)}{\partial(a_1, a_2, a_3)} = \frac{\partial(u_1, X_2, X_3)}{\partial(X_1, X_2, X_3)} \frac{\partial(X_1, X_2, X_3)}{\partial(a_1, a_2, a_3)} . \tag{2.8}$$

□

It follows from (2.7) that if the Eulerian velocity field is solenoidal or, in other words, if the fluid is incompressible:

$$\frac{\partial}{\partial x_l} u_l[x_i, t] = 0 , \tag{2.9}$$

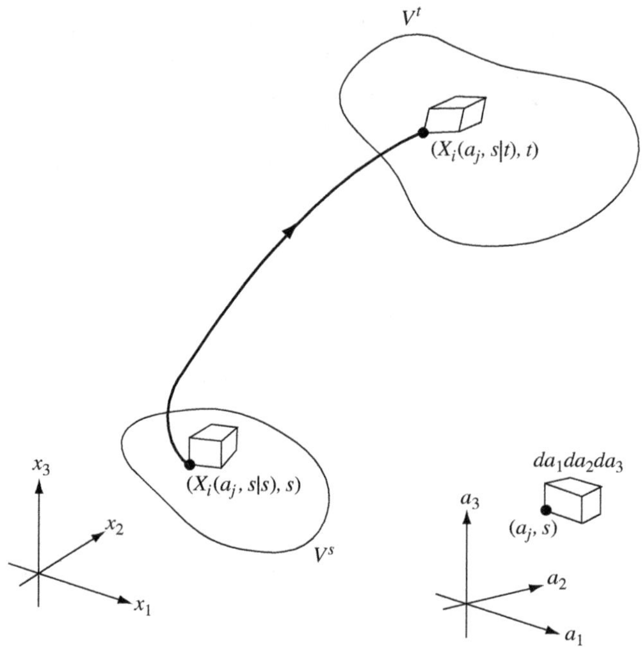

Figure 2.1 A parcel of fluid occupying the volume V^s at time s occupies the volume V^t at time t.

then for all t,

$$J_s^t = J_s^s . \tag{2.10}$$

So for an incompressible fluid,

$$I = \int_{V^s} \mathcal{U}(u_j(a_k, s|t))dV^s = \int_{V^t} \mathcal{U}(u_j[x_i, t])dV^t . \tag{2.11}$$

If the Eulerian velocity is statistically homogeneous then so is the Lagrangian velocity, and the expectation value of the integral on the left-hand side of (2.11) is simply the uniform value of the expectation of the integrand, multiplied by the volume V^s of the domain (the same notation will be used for both the domain and its volume). Such a simple statement cannot be immediately made about the integral on the right-hand side of (2.11), since the domain V^t is flow dependent. Yet the Eulerian flow being assumed homogeneous implies that the fluid occupies an unbounded region. The domain V^s may be chosen to be a sphere of very large radius, in which case V^t will be the same except for very small surface distortions. So with sufficient accuracy, after taking the expectation of I and dividing by V^s,

$$E\{\mathcal{U}(u_j(a_k, s|t))\} = E\{\mathcal{U}(u_j[x_i, t])\}. \tag{2.12}$$

In (2.12), $E\{\theta\}$ denotes the expectation value of θ, for any θ. That is,

$$E\{\theta\} = \int \theta P(\theta) d\theta, \tag{2.13}$$

where $P(\theta)$ is the probability distribution function or pdf for θ. In particular, let $\mathcal{U}(u_k) = \exp(i\lambda_j u_j)$; in that case $E\{\mathcal{U}(u_k)\}$ is the characteristic function of the random variable u_k, or Fourier transform of the probability distribution for u_k. Thus (2.12) implies that in a flow that is incompressible and homogenous, the Eulerian velocity at any point in space and at any time has the same pdf as the Lagrangian velocity for a single fluid particle at any time after release (Lumley, 1962).

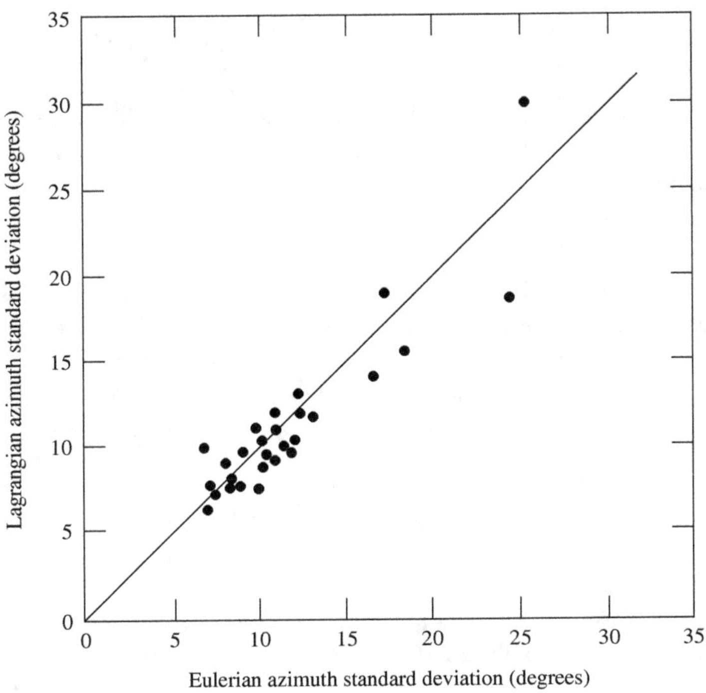

Figure 2.2 Equivalence of standard deviations of azimuthal Eulerian and Lagrangian velocity: winds over prairie grass. (Reviewed in Lumley and Panofsky, 1964, Section 4.5).

Are the standard deviations of Eulerian and Lagrangian velocity, for example, ever observed to be the same? Some data for wind over prairie grass are reviewed by Lumley and Panofsky (1964; Section 4.5). Their Figure 4.13 is reproduced here as Figure 2.2; it shows approximate equality of the azimuthal standard deviations.

2.2 Single-particle, two-time statistics

Consider a second moment of the Lagrangian velocity of a single particle, the factors being evaluated at different times:

$$S_{ij}(a_k, s|t_1, t_2) \equiv E\{u_i(a_k, s|t_1)u_j(a_k, s|t_2)\}. \tag{2.14}$$

In terms of Eulerian velocities and particle paths, (2.14) becomes

$$S_{ij}(a_k, s|t_1, t_2) \equiv E\{u_i[X_p(a_k, s|t_1), t_1]u_j[X_q(a_k, s|t_2), t_2]\}, \tag{2.15}$$

which may also be expressed as

$$S_{ij}(a_k, s|t_1, t_2) \equiv E\{u_i[X_p(a_k, s|t_1), t_1]u_j[X_q(X_p(a_k, s|t_1), t_1|t_2), t_2]\}, \tag{2.16}$$

that is, the position $X_p(a_k, s|t_1)$ of the particle at time t_1 is labeling the Lagrangian velocity of the particle at time t_2. Indeed, for simplicity make the assumption that $X_p(a_k, t_1|t_1) = a_p$, that is, particles are always labeled by position at some time.

Exercise 2.2 Prove that if the Eulerian velocity is statistically homogenous and statistically stationary, that is, if its multi-point pdf is independent of absolute positions and time, then the second moment for the single-particle Lagrangian velocity at two times after release is independent of the position and time of particle release, and is translation invariant with respect to the two times:

$$S_{ij}(a_k, s|t_1, t_2) = S_{ij}(t_1 - t_2). \tag{2.17}$$

Hints:

(i) Expand S as a Taylor series about t_1, in powers of $t_2 - t_1$. Observe that the spatial arguments of the coefficients are all evaluated at $X_p(a_k, s|t_1)$.
(ii) Expand $X_p(a_k, s|t_1)$ as a Taylor series about t_1, in powers of $s - t_1$. Substitute for the partial derivatives with respect to s, using the labeling theorem.

□

In the sense of (2.17), the single-particle, two-time velocity is statistically homogeneous in space and statistically stationary in time.

2.3 Two-particle, two-time statistics

All that can be said in general about the joint pdf for the two Eulerian velocities $u_i[x_k, t]$ and $u_j[y_l, w]$ on the one hand, and on the other about the joint pdf for the two Lagrangian velocities $u_i(a_k, s|t)$ and $u_j(b_l, s|w)$, is that they are not the same. It can be said that the space-time correlation of the Lagrangian velocities cannot be stationary, owing to the following simple consideration (Lumley and Panofsky, 1964): if $t-s$ and $w-s$ are both small, the correlation between the two velocities is determined by the initial separation $a_k - b_k$, which may be small; however, for large values of $t-s$ and $w-s$ such that $t-w$ is unaltered, the particles will in general have wandered far apart and so their velocities will have become uncorrelated.

2.4 The Eulerian–Lagrangian problem: path integrals

Lumley and Panofksy (1964) call the general problem of relating the statistics of the two formulations of fluid dynamics the *Eulerian–Lagrangian problem.* There is a general relationship in terms of ‘path integrals’, also known as functional integrals (e.g., Monin and Yaglom, 1975; Drummond, 1982). First, consider the position $X_i(a_j, s|t)$ of a fluid particle at time t, having been labeled by its position a_j at time s. For a single realization of a turbulent flow, and for any point x_i, the particle position X_i definitely is or is not within a Δx_i neighborhood of x_i. Let f be the fraction of realizations for which the particle is within such a neighborhood. Then the fractional density of particles in the neighborhood is approximately $f/(\Delta x_1 \Delta x_2 \Delta x_3)$, and for these realizations the density is zero elsewhere. That is to say, the probability distribution function or pdf P for the possible values x_i of the random variable $X_i(a_j, s|t)$, is

$$P(a_j, s|x_i, t) = E\{\delta(x_i - X_i(a_j, s|t))\}, \tag{2.18}$$

where $\delta(x_i)$ is the three-dimensional Dirac delta function.[1] It is in general more satisfactory, when many different pdfs are being discussed, to follow the convention (Feller, 1968) of denoting the pdf as $P_{X_i}(a_j, s|x_i, t)$ but such

[1] Also denoted $\delta(x_1)\delta(x_2)\delta(x_3)$.

elaboration will be cautiously avoided here. The position pdf defined in (2.18) may also be expressed as

$$P(a_j, s|x_i, t) = E\left\{\delta\left(x_i - a_i - \int_s^t u_i(a_j, s|r)dr\right)\right\}, \tag{2.19}$$

where $u_i(a_j, s|t)$ is the Lagrangian velocity. Discretizing in time so that

$$t^0 = s, \ldots, t^k = s + k\Delta t, \ldots, t^K = t; \quad x_i^0 = a_i, \ldots, x_i^K = x_i, \tag{2.20}$$

the position pdf is approximately

$$P(a_j, s|x_i, t) \cong E\left\{\delta\left(x_i^K - x_i^0 - \sum_{k=0}^{K-1} u_i(x_i^0, t^0|t^k)\Delta t\right)\right\}. \tag{2.21}$$

A simple substitution leads to

$$\begin{aligned} P(a_j, s|x_i, t) \cong E\Bigg\{ \int \delta\left(x_i^K - x_i^{K-1} - u_i[x_i^{K-1}, t^{K-1}]\Delta t\right) \\ \times \delta\left(x_i^{K-1} - x_i^0 - \sum_{k=0}^{K-2} u_i(x_i^0, t^0|t^k)\Delta t\right) dV^{K-1}\Bigg\}, \end{aligned} \tag{2.22}$$

where dV^{K-1} is the volume element in x_i^{K-1} space. Note the emergence of the Eulerian velocity $u_i[x_n, t] \equiv u_i(x_n, t|t)$, which satisfies $u_i[X_n(a_j, s|t), t] = u_i(a_j, s|t)$. Continuing with these substitutions, and rescaling the measure, leads to

$$\begin{aligned} P(a_j, s|x_i, t) \cong E\Bigg\{ \prod_{k=1}^{K-1} \int \frac{dV^k}{(\Delta t)^N} \delta\left(\frac{x_i^{k+1} - x_i^k}{\Delta t} - u_i[x_n^k, t^k]\right) \\ \times \delta\left(\frac{x_i^1 - x_i^0}{\Delta t} - u_i[x_i^0, t^0]\right)\Bigg\}, \end{aligned} \tag{2.23}$$

where N is the number of space dimensions (here, $N = 3$). So the Lagrangian pdf for the position of a fluid particle may be expressed in terms of the Eulerian velocity field $u_i[x_n, t]$. Accordingly, the expectation in (2.23) invokes the Eulerian velocity pdf:

$$\begin{aligned} P(a_j, s|x_i, t) \cong \prod_{k=1}^{K-1} \int \frac{dV^k}{(\Delta t)^N} \int dU^k \int dU^0 \delta\left(\frac{x_i^{k+1} - x_i^k}{\Delta t} - u_i^k\right) \delta\left(\frac{x_i^1 - x_i^0}{\Delta t} - u_i^0\right) \\ \times p^{\mathrm{Eul}}\left(t^0, \ldots, t^k, \ldots; x_n^0, \ldots, x_n^k, \ldots; u_i^0 \ldots, u_i^k, \ldots\right), \end{aligned} \tag{2.24}$$

where dU^k is the volume element in u_i^k space, and P^{Eul} is the joint pdf of the Eulerian velocities u_i^k at (x_n^k, t^k), for $k=0, \ldots, K-1$. Performing the integrations with respect to all the u_n^k yields

$$P(a_j, s|x_i, t) \cong \prod_{k=1}^{K-1} \int \frac{dV^k}{(\Delta t)^N} P^{\mathrm{Eul}} \left(t^0, \ldots, t^k, \ldots ; x_n^0, \ldots, x_n^k, \ldots ; \frac{x_n^1 - x_n^0}{\Delta t}, \ldots, \frac{x_n^{k+1} - x_n^k}{\Delta t}, \ldots \right). \tag{2.25}$$

In the limit, path integrals like (2.25) always attract gaudy notation, such as

$$P(a_j, s|x_i, t) = \int \mathfrak{D}[\Xi(r)] P^{\mathrm{Eul}} \left(r, \Xi, \dot{\Xi} \right), \tag{2.26}$$

where $\Xi(r)$ is *any* path from (a_j, s) to (x_i, t), not necessarily a fluid particle path.

The Eulerian velocity field in (2.23) is evaluated at (x_n^k, t^k). Could it be evaluated at $(x_n^k + \alpha \xi_n^k, t^k + \beta \Delta t)$ where $\xi_n^k = x_n^{k+1} - x_n^k$, preferably for $0 \le \alpha \le 1$ and $0 \le \beta \le 1$? Consider (2.19), with the expectation $E\{\quad\}$ deferred for simplicity. That is, consider the "micro" pdf, and just for a short time ($K=1$):

$$\pi(a_j, s|x_i, s+\Delta t) \equiv \delta \left(x_i - X_i(a_j, s|s+\Delta t) \right). \tag{2.27}$$

The path integral representation reduces to

$$\pi(a_j, s|x_i, s+\Delta t) = \delta(x_i - a_i - \Delta t u_i[a_j + \alpha \xi_j, s + \beta \Delta t]). \tag{2.28}$$

It follows, after allowing for the Jacobi determinant of the spatial perturbation of the Dirac argument in (2.28), that

$$\pi(a_j, s|x_i, s+\Delta t) \cong \left(1 + \alpha \Delta t \frac{\partial}{\partial x_i} u_i[a_j, s] - \Delta t u_i[a_j, s] \frac{\partial}{\partial x_i} \right) \times \pi(a_j, s|x_i, s) + O((\Delta t)^2). \tag{2.29}$$

On the other hand, the classical argument is that, since the particle must be *somewhere*,

$$\left(\frac{\partial}{\partial t} + \frac{\partial}{\partial x_i} u_i[x_j, t] + u_i[x_i, t] \frac{\partial}{\partial x_i} \right) \pi(a_j, s|x_i, t) = 0, \tag{2.30}$$

and so (2.29) is consistent with the Liouville equation (2.30) for all β if either $\alpha = -1$, which is remarkable, or, as Drummond (1982) points out, if the Eulerian velocity is solenoidal: $\partial u_i / \partial x_i = 0$, in which case the representation is independent of the choice of α. For a further discussion, see Schulman (1981).

It has been demonstrated above that the Lagrangian pdf for the position of a fluid particle is explicitly, if transcendentally, related to the Eulerian pdf for the fluid velocity. The pdf for any Lagrangian field is similarly related to the Eulerian velocity pdf. For example, let $P^{\text{Lag}}(a_j, s|U_i, t)$ be the pdf for the possible value U_i of the random Lagrangian velocity $u_i(a_j, s|t)$. Then, with dV^m being a volume element in x_m-space,

$$P^{\text{Lag}}(a_j, s|U_i, t) = E\left\{\delta\left(U_i - u_i(a_j, s|t)\right)\right\}$$
$$= E\left\{\int \delta\left(U_i - u_i[x_m, t]\right) \delta(x_m - X_m(a_j, s|t))\, dV^m\right\}, \quad (2.31)$$

which may be developed as a path integral for P^{Eul}. Note that it would be preferable here to denote the independent velocity variable as u_i and the Lagrangian velocity as $U_i(a_j, s|t)$, just as for the independent position x_i and particle path $X_i(a_j, s|t)$. However, the lower case/upper case convention cannot be sustained without causing other awkwardness elsewhere.

The Eulerian–Lagrangian relationship (2.26) is used by Bennett (1996) to test the Lagrangian position pdf P for Gaussianity, asymptotically for large t, when the Eulerian velocity pdf P^{Eul} is non-Gaussian and inhomogeneous. Non-Gaussianity is found for small to intermediate values of t. The findings for large t are inconclusive, owing to inadequate computing capacity and inefficient Monte Carlo algorithms for multidimensional integration. Also, the Monte Carlo trials in Bennett (1996) for particle displacement are not constrained to be consistent with a solenoidal Eulerian velocity field. Computing and Monte Carlo integration have since improved greatly, especially the latter (Ingber, 1993).

3
Lagrangian dynamics

3.1 Conservation of mass

The conservation law for mass has a compact Lagrangian form. At time t, let V^t be a fluid parcel or volume of particles of fixed identity. Fluid particles do not enter or leave the parcel. The fluid may be compressible and so the volume of the parcel may evolve in time, but the mass of the parcel is conserved. The expression of this principle in integral Eulerian form is:

$$\frac{d}{dt}\int_{V^t} \rho[x_i, t]dV^t = 0, \tag{3.1}$$

where ρ is the fluid density. At the labeling time s, the parcel occupies the domain W in labeling space. Transforming to labeling coordinates yields

$$\frac{d}{dt}\int_{W} \rho(a_j, s|t)J_s^t dW = 0, \tag{3.2}$$

where $J_s^t = \det(\partial X_i/\partial a_j)$ is the Jacobi determinant of the transformation $a_i \rightarrow x_i = X_i(a_j, s|t)$. The determinant is assumed always to be positive. The label a_i need not be the particle position at time s. That is, the labeling element of measure $dW = da_1 da_2 da_3$ in (3.2) need not have units of (length)^3. The validity of the change of variables is established in Exercise 1.4. The real spatial domain of the parcel at time s is V^s, and $dV^s = J_s^s dW$: see Figure 3.1.

The total time derivative d/dt commutes with the integral in labeling space, and the resulting integral vanishes for any parcel W, so

$$\frac{\partial}{\partial t}\{\rho(a_i, s|t)J_s^t(a_i, s|t)\} = 0. \tag{3.3}$$

This is the Lagrangian form of the law of conservation of mass.

It is instructive to compare the fluid parcel of (3.1) with the plume cross-sectional element of Lee and Chu (2003, Figure 3.9, p. 82). The latter parcel does not have fixed identity, owing to turbulent entrainment of fluid from

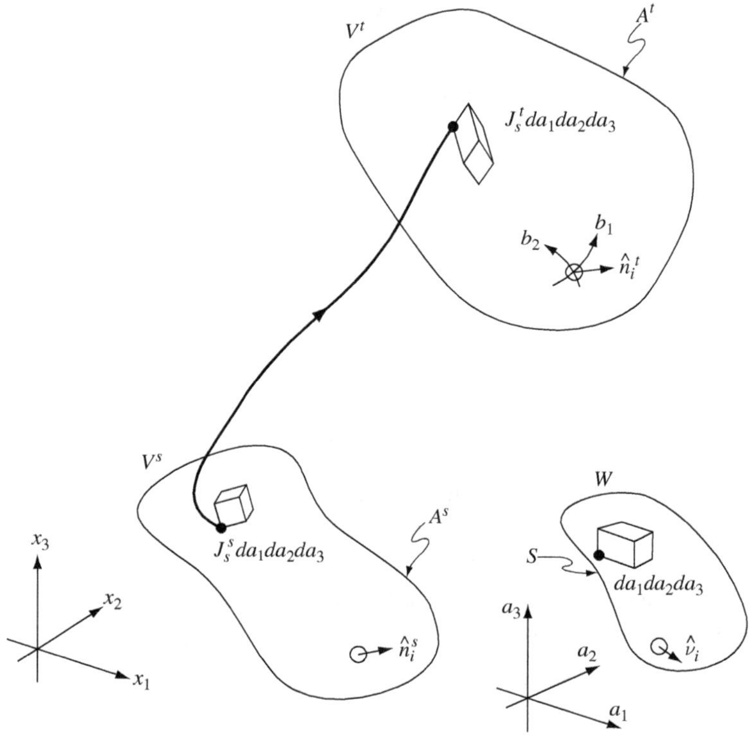

Figure 3.1 At time t, the volume element in a fluid parcel is $dV^t = J_s^t da_1 da_2 da_3$. The surface of the parcel is A^t, with outward unit normal $\hat{n}_i{}^t$ and surface coordinates (b_1, b_2).

ouside the plume. That is, Lee and Chu do not define the lateral boundary of the turbulent plume by a material surface. As a consequence, the fluid particles inside the plume element at any time may have been anywhere in the fluid at the initial instant. In any event, Lee and Chu suppress particle identity by integrating over the plume cross-section. Nevertheless, the elemental boundaries normal to the plume axis are material surfaces; in this sense the bulk equations of Lee and Chu are indeed "Lagrangian."

It is also instructive to examine Lee and Chu's splendid selection of photographs of turbulent jets and plumes.

3.2 Conservation of momentum

The stress tensor in a fluid is a state variable. The trace of the stress tensor, divided by the number of space dimensions, is the familiar state variable of pressure. Fluids are accelerated by spatial gradients of stresses. These gradients, being intrinsically Eulerian in nature, have awkward Lagrangian forms and so conservation of momentum is awkwardly expressed in Lagrangian form. For simplicity, only isotropic stresses will be admitted for now, that is, only pressure need be considered. It will be seen that certain steps in the construction of the Lagrangian form of conservation of momentum are simpler in the Eulerian form. But that is "cheating." So these steps will also be carried out in Lagrangian form, in the interest of gaining insight into Lagrangian fluid dynamics.

Let A^t be the surface of the fluid parcel at time t. Fluid particles do not cross this surface and so do not carry momentum into or out of the parcel, just as they do not carry mass in or out. Thus, in the absence of distributed forcing applied by an external agency, the rate of change of momentum in the parcel can owe only to surface stresses. Again, assume for now that there are no shear stresses, but only a normal stress or pressure p. Thus

$$\frac{d}{dt}\int_{V^t} \rho u_i dV^t = -\int_{A^t} p\hat{n}_i^t dA^t , \tag{3.4}$$

where $\hat{n}_i^t$ is the outward unit normal on A^t. As before,

$$\frac{d}{dt}\int_{V^t} \rho u_i dV^t = \int_W \frac{\partial}{\partial t}(\rho u_i J_s^t)\, dW = \int_W \rho J_s^t \frac{\partial u_i}{\partial t} dW , \tag{3.5}$$

after invoking conservation of mass (3.3). Applying the divergence theorem to the surface integral in (3.4) yields

$$-\int_{A^t} p\hat{n}_i^t dA^t = -\int_{V^t} \frac{\partial p}{\partial x_i} dV^t = -\int_W \frac{\partial A_k}{\partial x_i}\frac{\partial p}{\partial a_k} J_s^t dW , \tag{3.6}$$

where $a_k = A_k(x_i, t|s)$ labels a particle at time s given its position x_i at time t. The label a_k need not be the position at time s. If a_k *is* the particle position, then it is the case that $a_k = X_k(x_i, t|s)$. The parcel W is arbitrary, so

$$\rho\frac{\partial u_i}{\partial t} = -\frac{\partial A_k}{\partial x_i}\frac{\partial p}{\partial a_k} . \tag{3.7}$$

This is the Lagrangian form of the law of conservation of momentum for an inviscid fluid in the absence of external forces. At first inspection it might seem to be of mixed form, owing to the presence of the partial derivative of the labeling position A_k with respect to the Eulerian position x_i, but the resulting matrix-valued field can be prescribed with Lagrangian arguments. Nevertheless, it is revealing to multiply both sides of (3.7) by the inverse matrix, yielding the strictly Lagrangian form

$$\rho\frac{\partial X_i}{\partial a_k}\frac{\partial u_i}{\partial t} = -\frac{\partial p}{\partial a_k} . \tag{3.8}$$

Thus at time t, the strained Lagrangian acceleration is down the pressure gradient with respect to the labels.

Exercise 3.1 By taking the partial derivative of

$$a_j = A_j(x_m, t|s) = A_j(X_m(a_k, s|t), t|s) \tag{3.9}$$

with respect to a_k, find the arguments of the inverse of the matrix in (3.7). □

Exercise 3.2 Consider a matrix with elements C_{ij}. Prove Cramer's rule: the inverse matrix has elements D_{ij} given by

$$D_{ij} = \frac{1}{2C} \epsilon_{imn} \epsilon_{jlk} C_{lm} C_{kn} , \tag{3.10}$$

where $C = \det(C_{ij})$ and $\epsilon_{123} = 1$, etc. Hint: combine Examples 4 and 7 on page 15 of Jeffreys (1931). □

Aside: It is "cheating" to have manipulated (3.4) by invoking the divergence theorem in the Eulerian coordinates (that is, in x_i at time t). The spirit of this development requires that all such maneuvers be made in the Lagrangian coordinates a_i at time s. So let b_1, b_2 be two surface coordinates on A^t, such that x_i lies on the surface if and only if

$$x_i = \xi_i^t(b_1, b_2) ,$$

for some function ξ_i^t. See Figure 3.1. Then the directed surface element on A^t is

$$\hat{n}_i^t dA^t = \epsilon_{ilj} \frac{\partial \xi_l^t}{\partial b_1} \frac{\partial \xi_j^t}{\partial b_2} db_1 \, db_2 .$$

Now at time s, the label a_i lies on the parcel surface S in labeling space provided

$$a_i = \zeta_i(b_1, b_2) ,$$

for some function ζ_i. Note that ζ_i differs from ξ_i^s if the label a_i is not the particle position at time s. If at time t the point x_i lies on the particle path labeled by a_j, then

$$x_i = \xi_i^t(b_1, b_2) = X_i\big(\zeta_j(b_1, b_2), s|t\big) ,$$

and so

$$\hat{n}_i^t dA^t = \epsilon_{ilj} \frac{\partial X_l}{\partial a_p} \frac{\partial \zeta_p}{\partial b_1} \frac{\partial X_j}{\partial a_q} \frac{\partial \zeta_q}{\partial b_2} db_1 \, db_2 .$$

The directed surface element on S, that is

$$\hat{\nu}_i dS = \epsilon_{ilj} \frac{\partial \zeta_l}{\partial b_1} \frac{\partial \zeta_j}{\partial b_2} db_1 \, db_2 ,$$

is beginning to emerge through the fog. The outward unit normal $\hat{\nu}_i$ on S differs from $\hat{n}_i^s$ on A^s if the label a_i is not the particle position at time s. Reversing the l, j indices and renaming the p, q indices yields

$$\hat{n}_i^t dA^t = \frac{1}{2}(\epsilon_{ilj} - \epsilon_{ijl})\frac{\partial X_l}{\partial a_p}\frac{\partial X_j}{\partial a_q}\left(\frac{\partial \zeta_p}{\partial b_1}\frac{\partial \zeta_q}{\partial b_2} - \frac{\partial \zeta_p}{\partial b_2}\frac{\partial \zeta_q}{\partial b_1}\right) db_1\, db_2\,,$$

or, since $\epsilon_{\mu\nu\lambda}\epsilon_{\lambda\theta\phi} = \delta_{\mu\theta}\delta_{\nu\phi} - \delta_{\mu\phi}\delta_{\nu\theta}$, where μ, ν, etc. $= 1, 2,\ 3$:

$$\hat{n}_i^t dA^t = \frac{1}{2}(\epsilon_{ilj} - \epsilon_{ijl})\frac{\partial X_l}{\partial a_\mu}\frac{\partial X_j}{\partial a_\nu}\epsilon_{\mu\nu\lambda}\hat{\nu}_\lambda dS\,.$$

The divergence theorem may now be invoked in the labeling coordinates a_i:

$$-\int_{A^t} p\hat{n}_i^t dA^t = -\int_W \frac{\partial}{\partial a_\lambda}\left(\frac{1}{2}(\epsilon_{ilj} - \epsilon_{ijl})p\frac{\partial X_l}{\partial a_\mu}\frac{\partial X_j}{\partial a_\nu}\epsilon_{\mu\nu\lambda}\right) dW\,,$$

or

$$-\int_{A^t} p\hat{n}_i^t dA^t = -\int_W \frac{1}{2}(\epsilon_{ilj} - \epsilon_{ijl})\epsilon_{\mu\nu\lambda}\frac{\partial p}{\partial a_\lambda}\frac{\partial X_l}{\partial a_\mu}\frac{\partial X_j}{\partial a_\nu} dW\,.$$

For example, if $i = 1$ the conclusion from (3.5) is

$$\rho J_s^t \frac{\partial u_1}{\partial t} = -\frac{\partial(p, X_2, X_3)}{\partial(a_1, a_2, a_3)}\,,$$

or, since $J_s^t J_t^s = 1$,

$$\rho\frac{\partial u_1}{\partial t} = -\frac{\partial(p, X_2, X_3)}{\partial(x_1, x_2, x_3)} = -\frac{\partial p}{\partial x_1} = -\frac{\partial p}{\partial a_k}\frac{\partial A_k}{\partial x_1}\,,$$

as before.

Alternatively, the surface coordinates b_1, b_2 may be constructed first on S, and then mapped onto A^t using particle paths. In either construction, it is assumed that the surface S in label space maps onto the parcel surface A^t, as will be the case if the map is nonsingular.

The Lagrangian form of conservation of momentum, derived without cheating in the preceding aside, that is:

$$\rho J_s^t \frac{\partial u_1}{\partial t} = \rho J_s^t \frac{\partial^2 X_1}{\partial t^2} = -\frac{\partial(p, X_2, X_3)}{\partial(a_1, a_2, a_3)}\,, \tag{3.11}$$

etc., reveals that there is an acceleration in any direction, so long as the pressure field is not functionally dependent solely upon the orthogonal directions. The Eulerian form of conservation of momentum:

$$\frac{\partial}{\partial t}u_i[x_j, t] + u_k[x_j, t]\frac{\partial}{\partial x_k}u_i[x_j, t] = -\frac{1}{\rho}\frac{\partial}{\partial x_i}p[x_j, t]\,, \tag{3.12}$$

expresses the same principle but lacks the geometrical framework of the Lagrangian form (3.11).

Exercise 3.3 Assume that an external force F_i per unit mass is applied to the fluid. Show that (3.8) becomes

$$\rho\frac{\partial X_i}{\partial a_k}\left(\frac{\partial u_i}{\partial t}-F_i\right)=-\frac{\partial p}{\partial a_k}. \tag{3.13}$$

□

3.3 Conservation of energy

The temperature T, entropy η, internal energy $\mathfrak{E}$, pressure p and density ρ of a fluid are state variables related by the combined first and second laws of thermodynamics:

$$Td\eta=d\mathfrak{E}+pd\left(\rho^{-1}\right). \tag{3.14}$$

The Lagrangian form of the internal energy equation for isentropic motion follows immediately:

$$T\frac{\partial\eta}{\partial t}=\frac{\partial\mathfrak{E}}{\partial t}+p\frac{\partial\left(\rho^{-1}\right)}{\partial t}=0. \tag{3.15}$$

Assuming that there are only two independent state variables, for example entropy η and density ρ, and assuming again that entropy is conserved, it follows that Lagrangian changes in pressure are related to those in density by

$$\frac{\partial p}{\partial t}=\left(\frac{\partial p}{\partial\rho}\right)_\eta\frac{\partial\rho}{\partial t}. \tag{3.16}$$

The coefficient in (3.16) is a state variable having the dimensions of the square of a speed; this speed will be subsequently identified as the speed of sound c. Both (3.15) and (3.16) express conservation of entropy, and both require an equation of state, such as $\mathfrak{E}=\mathfrak{E}(p,\rho)$ or $p=p(\eta,\rho)$.

Alternatively one may argue that, if neither an external field nor a heat source is present, the total energy in a fluid parcel consists of internal energy plus kinetic energy; their sum changes only as a result of work done by the fluid against the pressure on the surface of the parcel:

$$\frac{d}{dt}\int_{V^t}\rho\left(\mathfrak{E}+\frac{1}{2}u_ju_j\right)dV^t=-\int_{A^t}pu_k\hat{n}_k^t dA^t, \tag{3.17}$$

which, after invoking conservation of mass and momentum, yields

$$\rho \frac{\partial \mathfrak{E}}{\partial t} = -p \frac{\partial u_k}{\partial x_k} . \tag{3.18}$$

This mixed Eulerian–Lagrangian expression may be converted to a purely Lagrangian expression by transforming the flow divergence to Lagrangian form. The rate of change of total volume of a fluid parcel equals the rate at which volume is swept by its moving boundary:

$$\frac{dV^t}{dt} = \int_{A^t} u_k \hat{n}_k^t \, dA^t , \tag{3.19}$$

yielding

$$\frac{\partial J_s^t}{\partial t} = \frac{\partial u_k}{\partial x_k} J_s^t . \tag{3.20}$$

Note that this differential equation is mixed: the time derivative is Lagrangian while the space derivatives are Eulerian. However the conservation law (3.18) for internal energy can take the purely Lagrangian form

$$\rho J_s^t \frac{\partial \mathfrak{E}}{\partial t} + p \frac{\partial J_s^t}{\partial t} = 0 . \tag{3.21}$$

Invoking conservation of mass leads again to

$$\frac{\partial \mathfrak{E}}{\partial t} + p \frac{\partial (\rho^{-1})}{\partial t} = 0 . \tag{3.22}$$

The Lagrangian rate of change of the kinetic energy per unit mass $K = \frac{1}{2} u_i u_i$ follows from the momentum equation:

$$\rho J_s^t \frac{\partial K}{\partial t} = -J_s^t \frac{\partial}{\partial x_j} (u_j p) + p \frac{\partial J_s^t}{\partial t} . \tag{3.23}$$

Combining (3.21) and (3.23), the Lagrangian rate of change of total energy becomes

$$\frac{\partial}{\partial t} \{ \rho J_s^t (K + \mathfrak{E}) \} = -J_s^t \frac{\partial A_j}{\partial x_k} \frac{\partial}{\partial a_j} (u_k p) . \tag{3.24}$$

Note:

(i) $\partial A_j / \partial x_k$ is the matrix inverse of the Lagrangian field $\partial X_j / \partial a_k (a_l, s|t)$, so (3.24) is purely Lagrangian.

(ii) The Lagrangian conservation law for total energy is awkward, owing to the need to express the intrinsically Eulerian divergence of the rate of work done locally by pressure.

3.4 Variational principle

Assume that the thermodynamic state of the fluid is defined by two variables, for example the density ρ and the entropy η. In particular, the internal energy $\mathfrak{E}$ is a function of ρ and η alone, and the combined first and second laws (3.14) implies

$$\left(\frac{\partial \mathfrak{E}}{\partial \rho}\right)_\eta = \frac{p}{\rho^2}, \tag{3.25}$$

$$\left(\frac{\partial \mathfrak{E}}{\partial \eta}\right)_\rho = T. \tag{3.26}$$

Exercise 3.4 Show that conservation of momentum may be derived as the extremal condition for the Lagrange functional (Eckart, 1960; Seliger and Witham, 1968)

$$L = \int_s^{t_1} \int_W \Lambda dt dW \tag{3.27}$$

given some fixed time t_1, where the Lagrange density Λ is

$$\Lambda = \Lambda(\frac{\partial X_i}{\partial t}, \frac{\partial X_i}{\partial a_j}, \rho, \eta) = \rho J_s^t \left\{ \frac{1}{2} \frac{\partial X_j}{\partial t} \frac{\partial X_j}{\partial t} - \mathfrak{E}(\rho, \eta) \right\}, \tag{3.28}$$

subject to the constraints of conservation of mass (3.3) and entropy (3.15). That is, show the constrained Euler–Lagrange equation to be

$$\rho J_s^t \frac{\partial^2 X_j}{\partial t^2} = -\frac{\partial p}{\partial a_k} \frac{\partial J_s^t}{\partial \left(\frac{\partial X_j}{\partial a_k}\right)}, \tag{3.29}$$

which is the same as (3.11) when $j = 1$, etc.
Hints:

(i) assume a large domain, with motion vanishing in the far field;
(ii) express the mass constraint in the form $\rho(a_i, s|t)\ J(a_i, s|t) = \rho(a_i, s|s)\ J(a_i, s|s)$, vary neither the initial density nor the initial Jacobi determinant;
(iii) express the entropy constraint in the form $\eta(a_i, s|t) = \eta(a_i, s|s)$; do not vary the initial entropy.

Now assume that the functions X_i, ρ, η obey the equations of motion, and so the value of the integral L is extreme. The value will be unaffected by any substitution of the form $t = \tau(r)$ for the independent variable of time. For simplicity, assume that the substitution leaves the terminals unchanged:

$s = \tau(s)$, $t_1 = \tau(t_1)$. Establish, as a consequence of Λ having no explicit dependence upon t, the conservation of energy:

$$\frac{d}{dt}\int_W \rho J_s^t \left(\mathfrak{E} + \frac{1}{2}\frac{\partial X_i}{\partial t}\frac{\partial X_i}{\partial t}\right) dW = 0 \, . \tag{3.30}$$

Hint: Let $Y_i(a_j, s|r) \equiv X_i(a_j, s|\tau(r))$, etc.; consider arbitrary variations of Y_i, etc. and τ. Compare (3.30) with (3.17), and recall the assumption here that the motion vanishes in the far field.

Investigate the conservation of the curl, in labeling coordinates, of the momentum per unit measure in labeling space. □.

3.5 Bernoulli's theorem

Fluid motion does tend to be time dependent, yet the principle of conservation of energy for steady flow is the foundation of engineering hydrodynamics. It is simpler to derive Bernoulli's theorem directly for steady flow, rather than by manipulating the time-dependent energy equation (3.21).

Invoking the law of conservation of momentum (3.7), the Lagrangian rate of change of kinetic energy is

$$\frac{\partial}{\partial t}\frac{1}{2}u_k u_k = -\frac{1}{\rho}u_i \frac{\partial A_j}{\partial x_i}\frac{\partial p}{\partial a_j} \, . \tag{3.31}$$

In particular, u_j denotes $u_j(a_l|t-s)$ in the above expression. Indeed, for steady flow $X_j(a_l, s|t) = X_j(a_l|t-s)$, and so

$$u_j(a_l, s|t) = u_j(a_l|t-s) = \frac{\partial}{\partial t}X_j(a_l|t-s) = -\frac{\partial}{\partial s}X_j(a_l|t-s) \, . \tag{3.32}$$

By the purely kinematical labeling theorem,

$$\frac{\partial}{\partial s}X_j(a_l|t-s) = -u_k[a_l]\frac{\partial}{\partial a_k}X_j(a_l|t-s) \, , \tag{3.33}$$

and so

$$\begin{aligned}\frac{\partial}{\partial t}\left(\frac{1}{2}u_k u_k\right)(a_l|t-s) &= -\frac{1}{\rho}u_n[a_l]\left(\frac{\partial X_i}{\partial a_n}\frac{\partial A_j}{\partial x_i}\frac{\partial p}{\partial a_j}\right)(a_l|t-s) \\ &= -\frac{1}{\rho}u_j[a_l]\frac{\partial}{\partial a_j}p(a_l|t-s) \, . \end{aligned} \tag{3.34}$$

Using the labeling theorem again,

$$\frac{\partial}{\partial t}\left(\frac{1}{2}u_k u_k\right)(a_l|t-s) = \frac{1}{\rho}\frac{\partial}{\partial s}p(a_l|t-s) = -\frac{1}{\rho}\frac{\partial}{\partial t}p(a_l|t-s) \, . \tag{3.35}$$

Assume that the fluid is barotropic: $p = p(\rho)$, and define the state variable of enthalpy $h = \int \rho^{-1} dp$.

Bernoulli's theorem

Along any streamline,

$$b = \frac{1}{2} u_k u_k + h \tag{3.36}$$

is a constant. ■

Notes

(i) The flow may be compressible, provided it is barotropic.
(ii) The labeling theorem, or explicit recognition of conserved particle identity, is invoked. It has been assumed that particle labels are release positions.

Exercise 3.5 Derive Bernoulli's theorem for steady flow from (3.24). □

3.6 Kelvin's theorem

The momentum equation in the presence of a force F_i per unit mass is

$$\rho \frac{\partial u_i}{\partial t} = -\frac{\partial A_j}{\partial x_i} \frac{\partial p}{\partial a_j} + \rho F_i . \tag{3.37}$$

Multiplying by the Jacobi or strain matrix yields

$$\frac{\partial X_i}{\partial a_k} \frac{\partial u_i}{\partial t} = -\frac{\partial h}{\partial a_k} + \frac{\partial X_i}{\partial a_k} F_i \tag{3.38}$$

where the fluid has been assumed barotropic: $p = p(\rho)$, and the enthalpy is again defined by $dh = dp/\rho$. Taking the curl of (3.38) yields

$$\epsilon_{jlk} \frac{\partial}{\partial a_l} \left(\frac{\partial X_i}{\partial a_k} \frac{\partial u_i}{\partial t} \right) = \epsilon_{jlk} \frac{\partial}{\partial a_l} \left(\frac{\partial X_i}{\partial a_k} F_i \right) . \tag{3.39}$$

Some manipulations of the left-hand side of (3.39) yield

$$\frac{\partial}{\partial t} \left(\epsilon_{jlk} \frac{\partial X_i}{\partial a_k} \frac{\partial u_i}{\partial a_l} \right) = \epsilon_{jlk} \frac{\partial}{\partial a_l} \left(\frac{\partial X_i}{\partial a_k} F_i \right) . \tag{3.40}$$

Integrating over the area of a surface S in label space yields, after further shuffling,

$$\frac{d}{dt} \int_S \epsilon_{jlk} \frac{\partial}{\partial a_l} \left(\frac{\partial X_i}{\partial a_k} u_i \right) dS = \int_S \epsilon_{jlk} \frac{\partial}{\partial a_l} \left(\frac{\partial X_i}{\partial a_k} F_i \right) dS . \tag{3.41}$$

Now suppose that S is an open surface in label space bounded by a closed circuit D. The corresponding open surface A^s in real space is bounded by a closed circuit C^s. At time t, the open surface is A^t bounded by a closed circuit C^t: see Figure 3.2. Then by Stokes' theorem applied in label space,

$$\frac{d}{dt}\oint_D \frac{\partial X_i}{\partial a_k} u_i \hat{\sigma}_k dD = \oint_D F_i \frac{\partial X_i}{\partial a_k} \hat{\sigma}_k dD, \tag{3.42}$$

where $\hat{\sigma}_k$ is the unit tangent on the label circuit D, and dD is an element of arc measure along that circuit. But it is easily seen that

$$\frac{\partial X_i}{\partial a_k} \hat{\sigma}_k dD = \hat{\tau}_i^t dC^t, \tag{3.43}$$

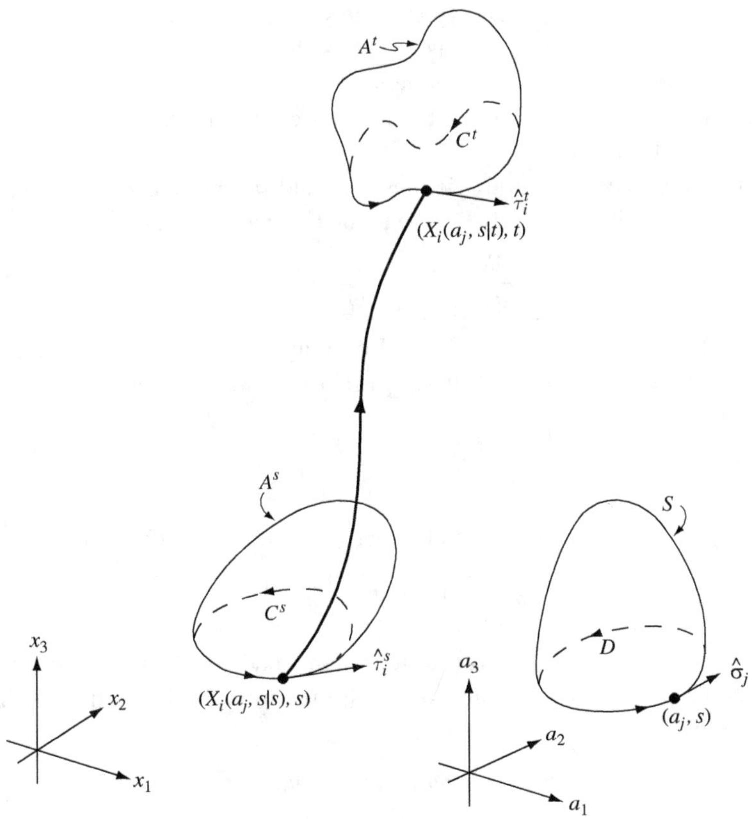

Figure 3.2 At time t, fluid particles lie on an open surface A^t, bounded by a closed circuit C^t with unit tangent $\hat{\tau}_i^t$.

where $\hat{\tau}_i^t$ is the unit tangent on the circuit C^t, and dC^t is an element of arc length along that circuit. The conclusion is the Lagrangian and classical form of Kelvin's theorem, for the rate of change of circulation around a closed circuit moving with the flow (Eckart, 1960):

Kelvin's theorem

$$\frac{d}{dt}\oint_{C^t} u_i \hat{\tau}_i^t \, dC^t = \oint_{C^t} F_i \hat{\tau}_i^t \, dC^t . \tag{3.44}$$

■

3.7 Cauchy–Weber integrals

3.7.1 First integrals

The preceding derivation of Kelvin's theorem commences with the application of the curl operator, after premultiplying the momentum equations by the strain or Jacobi matrix. This suggests that there are other Lagrangian fields to consider besides the Lagrangian expression of the Eulerian vorticity. Indeed, the inviscid Lagrangian momentum equation admits a first integral with respect to time.

In the presence of a barotropic pressure field $p = p(\rho)$, and a conservative body force F_i with potential $\mathcal{F}$, the strained momentum equation is

$$\frac{\partial X_i}{\partial a_j}\frac{\partial u_i}{\partial t} = -\frac{\partial}{\partial a_j}(h + \mathcal{F}) \tag{3.45}$$

where $dh = dp/\rho$ and $F_i = -\partial\mathcal{F}/\partial x_i$. Recalling that $u_i = \partial X_i/\partial t$, some simple manipulations (Lamb, 1932, Articles 13–15) yield the Cauchy–Weber integral relation

$$\frac{\partial X_i}{\partial a_j} u_i = -\frac{\partial \varphi}{\partial a_j} + v_j , \tag{3.46}$$

where the Cauchy–Weber integral scalar $\varphi = \varphi(a_i, s|t)$ is defined by

$$\varphi = \int_s^t \left(h + \mathcal{F} - \frac{1}{2} u_i u_i \right) dr \tag{3.47}$$

and the vector $v_j = v_j(a_k, s)$ is an arbitrary function of the labeling coordinates. If a_i is the particle position at time $t = s$, then v_j is the Eulerian velocity at time $t = s$:

$$v_i(a_j, s) = u_i(a_j, s|s) = u_i[a_j, s] . \tag{3.48}$$

The momentum-conservation law (3.46) reveals that the change in the strained velocity from time s to time t is minus the gradient of a scalar field φ with respect to the labels.

The time integrated law may be derived in another way. Taking the curl of the strained momentum equation (3.45) yields

$$\epsilon_{ilj}\frac{\partial}{\partial a_l}\frac{\partial}{\partial t}\left(\frac{\partial X_k}{\partial a_j}u_k\right)=0. \tag{3.49}$$

Simple manipulations yield

$$\frac{\partial}{\partial t}(\varpi_i)=0, \tag{3.50}$$

where

$$\varpi_i\equiv\epsilon_{ilj}\frac{\partial u_k}{\partial a_l}\frac{\partial X_k}{\partial a_j} \tag{3.51}$$

is therefore independent of the time since release:

$$\varpi_i(a_m, s|t)=\epsilon_{ijk}\frac{\partial}{\partial a_j}u_k[a_m, s], \tag{3.52}$$

assuming solely for simplicity that $X_i(a_j, s|s)=a_i$. By construction, ϖ_i is solenoidal:

$$\frac{\partial \varpi_i}{\partial a_i}=0. \tag{3.53}$$

Evidently ϖ_1, for example, is the circulation at the release time s around an infinitesimal circuit of fluid particle in label space, which circuit bounds the infinitesimal surface defined by the differentials da_2, da_3. But the body forces have been assumed conservative, so by Kelvin's theorem (3.44), ϖ_1 is also the value of the circulation at time t around the circuit defined by the same fluid particles. The surface bounded by the circuit at time t may not be orthogonal to the x_1 direction, even though a_1 may coincide with x_1 at time $t=s$. It will be convenient to name ϖ_i the *Cauchy invariant.*

Exercise 3.6

(i) Relate the two developments of the Cauchy–Weber integrals.

(ii) Assume that the particles are labeled by their release position, and that the fluid is at rest at that time. Show that the subsequent velocity field is the negative of the Eulerian gradient of the Cauchy–Weber integral scalar. Express the field arguments in detail.

□

Note that if $\varpi_i = 0$, then

$$\frac{\partial X_i}{\partial a_j} u_i = \frac{\partial \vartheta}{\partial a_j} \tag{3.54}$$

for some scalar field $\vartheta = \vartheta(a_j, s|t)$, and so

$$u_i = \frac{\partial A_j}{\partial x_i} \frac{\partial \vartheta}{\partial a_j} = \frac{\partial \vartheta}{\partial x_i} . \tag{3.55}$$

That is, the Eulerian vorticity vanishes: $\omega_i = 0$. It may be shown with a similar argument that if $\omega_i = 0$ then $\varpi_i = 0$. Thus, flow may be described unambiguously as either irrotational or rotational. Yet in general the Cauchy invariant ϖ_i is *not* the Eulerian vorticity ω_i. The latter is

$$\omega_i = \epsilon_{ilj} \frac{\partial u_j}{\partial x_l} = \epsilon_{ilj} \frac{\partial u_j}{\partial a_k} \frac{\partial A_k}{\partial x_l} . \tag{3.56}$$

Yes, (3.52) states that $\varpi_i(a_j, s|t) = \omega_i[a_j, s]$, under certain conditions, but these are not the values of two fields at the same point in space and at the same moment in time.

There are formal relationships between ϖ_i and ω_i.

Exercise 3.7 Show that

$$\varpi_i = J_s^t \frac{\partial A_i}{\partial x_j} \omega_j . \tag{3.57}$$

Hint (Jeffreys, 1931):

$$\epsilon_{lmn} J_s^t = \epsilon_{ijk} \frac{\partial X_l}{\partial a_i} \frac{\partial X_m}{\partial a_j} \frac{\partial X_n}{\partial a_k} = \epsilon_{ijk} \frac{\partial X_i}{\partial a_l} \frac{\partial X_j}{\partial a_m} \frac{\partial X_k}{\partial a_n} . \tag{3.58}$$

Explain the meaning of each symbol; in particular state the arguments of each term, and the relationships between the sets of arguments. Conclude that

$$\omega_i = (J_s^t)^{-1} \frac{\partial X_i}{\partial a_j} \varpi_j . \tag{3.59}$$

Are these linear relationships? Are they even explicit? □

Exercise 3.8 Show that for planar flow,

$$\varpi = J_s^t \omega . \tag{3.60}$$

□

Exercise 3.9 Return to the first development of the Cauchy–Weber integrals, at the stage where

$$\frac{\partial X_i}{\partial a_j}\frac{\partial u_i}{\partial t} = -\frac{\partial}{\partial a_j}(h+\mathcal{F}). \tag{3.61}$$

Assuming irrotational flow, show that

$$\frac{\partial}{\partial a_k}\left[h+\mathcal{F}+\frac{\partial \chi}{\partial t}-\frac{1}{2}u_j u_j\right]=0 \tag{3.62}$$

where

$$u_l\frac{\partial X_l}{\partial a_k} = \frac{\partial \chi}{\partial a_k} \tag{3.63}$$

for some scalar field $\chi = \chi(a_i, s|t)$. Reconcile this with the Eulerian form of Bernoulli's theorem. □

Exercise 3.10 Again assume irrotationally forced, barotopic inviscid flow. Derive from (3.50) and (3.57) the Eulerian form of the conservation law for the Eulerian vorticity $\omega_i = \omega_i[x_j, t]$:

$$\frac{\partial \omega_i}{\partial t} + u_j\frac{\partial \omega_i}{\partial x_j} + \omega_i\frac{\partial u_k}{\partial x_k} - \omega_k\frac{\partial u_i}{\partial x_k} = 0, \tag{3.64}$$

where the partial derivate with respect to time is Eulerian, that is, x_j is fixed. Hint: The Eulerian flow divergence is the Lagrangian logarithmic time derivative of the Jacobi determinant. □

3.7.2 Matrix formulation

The Cauchy invariant has inspired a "matrix" formulation of Lagrangian fluid dynamics (Yakubovich and Zenkovich, 2001). The Jacobi matrix or strain matrix $\partial X_i/\partial a_j$ is denoted J_{ij}. The invariant defined in (3.51) is expressed as the matrix $S_{ij} = \epsilon_{ijk}\varpi_k$. Then in fact (3.51) becomes

$$S_{ij} = \frac{\partial J_{ki}}{\partial t}J_{kj} - J_{ki}\frac{\partial J_{kj}}{\partial t}. \tag{3.65}$$

The matrix J_{ij} must satisfy the consistency conditions

$$\frac{\partial J_{ij}}{\partial a_k} = \frac{\partial J_{ik}}{\partial a_j}. \tag{3.66}$$

For incompressible flow, $J_s^t = \det(J_{ij})$ is independent of time t:

$$\det\left(J_{ij}(a_k, s|t)\right) = \det\left(J_{ij}(a_k, s|s)\right). \tag{3.67}$$

Exercise 3.11 The matrix notation for Eulerian vorticity is $W_{ij}=\epsilon_{ijk}\omega_k$. Show that

$$J_{li}W_{lk}J_{kj}=S_{ij}\,, \tag{3.68}$$

with solution

$$\omega_i=(J^t_s)^{-1}J_{ij}\varpi_j\,. \tag{3.69}$$

This is the Cauchy form of the vorticity equation (Batchelor, 1973), extended by Yakubovich and Zenkovich (2001) to general Lagrangian coordinates for which J^t_s is not necessarily identically equal to unity. That is, the flow need not be incompressible and the particle label a_i need not be the release position. □

Exercise 3.12 Subscripted notation fails us at this point. Let J be the Jacobi or strain matrix with elements J_{ij}. Let A be the skew-symmetric matrix with elements $A_{lm}=\epsilon_{lmn}A_n$, where the A_n are constants independent of all the Lagrangian coordinates $(a_i,\,s|t)$. Show (Yakubovich and Zenkovich, 2001) that

$$\mathsf{J}=e^{\mathsf{A}(t-s)}\mathsf{J}_s \tag{3.70}$$

is a solution of the matrix forms of the equations of Lagrangian incompressible fluid dynamics, provided $\mathsf{J}_s=\mathsf{J}(a_i,\,s|s)$ satisfies the consistency conditions (3.66). Find the matrix S which has components S_{ij}. Show that this solution corresponds to rigid rotation of the fluid.
Hint: exponentiated skew matrices are unimodular. □

3.7.3 Cauchy–Weber integrals and Clebsch potentials

Recall the Cauchy–Weber integral (3.46):

$$\frac{\partial X_i}{\partial a_i}u_i=-\frac{\partial\varphi}{\partial a_j}+v_j\,, \tag{3.71}$$

where the scalar $\varphi(a_k,\,s|t)$ as defined by (3.47) vanishes at $t=s$, and $v_j(a_k,\,s)$ is independent of t. Recall the inessential but simplifying assumption that $X_i=a_i$ at $t=s$, and so

$$v_j(a_k,\,s)=u_j(a_k,\,s|s)=u_j[a_k,\,s]\,. \tag{3.72}$$

Lamb (1932, Art. 167) shows that there are scalar fields ξ,μ,λ where $\xi\;=\;\xi(a_k,s)$, etc., such that

$$v_j=-\frac{\partial\xi}{\partial a_j}+\lambda\frac{\partial\mu}{\partial a_j}\,. \tag{3.73}$$

It follows that the curl of (3.71) with respect to a_j is

$$\varpi_j = \epsilon_{jlk} \frac{\partial \lambda}{\partial a_l} \frac{\partial \mu}{\partial a_k}, \tag{3.74}$$

which should be compared with (3.51), (3.52). Thus lines tangential to ϖ_j coincide with the intersections of the sheets of constant μ and λ.

The link is completed by noting that the Lagrangian velocity u_i in the Cauchy–Weber integral (3.46) is related to these potentials via

$$\frac{\partial X_i}{\partial a_j} u_i = -\frac{\partial H}{\partial a_j} + \lambda \frac{\partial \mu}{\partial a_j} \tag{3.75}$$

where $H = \varphi + \xi$. Indeed,

$$\frac{\partial H}{\partial t} = h + \mathcal{F} - \frac{1}{2} u_i u_i. \tag{3.76}$$

The scalars $H(a_j, s|t)$, $\lambda(a_j, s)$ and $\mu(a_j, s)$ are the Clebsch potentials. They constitute the canonical variables in the Hamiltonian formulation of fluid dynamics (Zakharov and Kuznetsov, 1997).

3.8 Potential flow and a Riemannian metric

If the velocity is irrotational, that is if the vorticity vanishes, $\omega_i = 0$, then there is an Eulerian velocity potential χ such that

$$u_i = \frac{\partial \chi}{\partial x_i}. \tag{3.77}$$

If the velocity is also solenoidal, then the velocity potential obeys Laplace's equation:

$$\frac{\partial u_i}{\partial x_i} = \frac{\partial^2 \chi}{\partial x_i \partial x_i} = 0. \tag{3.78}$$

The velocity may therefore be constructed from boundary conditions using images, Green's functions, numerical methods, etc.

There is a Lagrangian construction of Laplace's equation. If the flow is incompressible, then the volume of a fluid parcel does not change in time. Reviewing the analysis in Section 3.1 leads in this case to

$$\frac{\partial J_s^t}{\partial t} = 0 \tag{3.79}$$

where $J_s^t = \det\left(J_{ij}\right)$ is the determinant of the Jacobi matrix $J_{ij} = \partial X_i/\partial a_j$. The incompressibility condition (3.79) may be expressed as (Jeffreys, 1931)

$$\frac{\partial}{\partial t}\left(\epsilon_{ijk}\epsilon_{lmn}\frac{\partial X_i}{\partial a_l}\frac{\partial X_j}{\partial a_m}\frac{\partial X_k}{\partial a_n}\right) = 0 . \tag{3.80}$$

If the flow is irrotational, that is if $\varpi_i = 0$, then there is a Lagrangian velocity potential $\varkappa$ such that

$$J_{ji}u_j = \frac{\partial \varkappa}{\partial a_i} . \tag{3.81}$$

It is convenient here to introduce the notation K_{ij} for an element of the inverse of the Jacobi matrix with elements J_{ij}. In terms of the position-to-label transformation $a_i = A_i(x_j, t|s)$ of Section 3.2,

$$K_{ij} = \frac{\partial A_i}{\partial x_j} . \tag{3.82}$$

Exercise 3.13 Use Cramer's rule (see Exercise 3.2) to show that

$$\sqrt{g}^{-1}\frac{\partial}{\partial a_i}\left(\sqrt{g}g^{ij}\frac{\partial \varkappa}{\partial a_j}\right) = 0 , \tag{3.83}$$

where $g^{ij} \equiv K_{il}K_{jl}$ is the matrix inverse of $g_{ij} \equiv J_{li}J_{lj}$.

Hint:

$$\frac{\partial}{\partial a_i}(J_s^t K_{ij}) = 0 . \tag{3.84}$$

□

It follows directly (Monin, 1962; Nakahara, 1990) that $\varkappa$ is harmonic in the labeling coordinates, with respect to the Riemannian metric

$$dC^2 = da_i g_{ij} da_j = da_i \frac{\partial X_l}{\partial a_i}\frac{\partial X_l}{\partial a_j} da_j = dX_l dX_l , \tag{3.85}$$

of course. It may be laboriously verified that the Riemannian curvature for the metric g_{ij} vanishes identically in label space. That is, label space is flat in this metric. The property is inherited from the flatness of Eulerian space in the Euclidean metric, which flatness is also shown explicitly in (3.85).

3.9 Boundary conditions

3.9.1 Rigid boundaries

Let A be a surface which is the locus of points satisfying

$$A(x_i)=0. \tag{3.86}$$

As before, no notational distinction is made here between the set of points A and the function A which defines it. The unit outward normal to the surface is parallel to the gradient of A. The surface is *rigid* if the flow at the surface is always tangential:

$$u_k(a_j, s|t)\frac{\partial A}{\partial x_k}(X_i(a_j, s|t))=0, \tag{3.87}$$

whenever $x_i=X_i(a_j, s|t)$ lies on the surface. But $u_i(a_j, s|t)=\frac{\partial X_i}{\partial t}(a_j, s|t)$, and so

$$\frac{\partial}{\partial t}A(X_i(a_j, s|t))=0. \tag{3.88}$$

If a point lies on such a surface at the labeling time $t=s$, then $A(X_i(a_j, s|s))=0$ and so $A(X_i(a_j, s|t))=0$. That is, the point must be on the surface at all other times.

The rigidity condition (3.87) only has meaning at boundary points where the boundary function is differentiable; that is, corners and cusps are excluded. The inference (3.88) is valid only if the particle paths are smooth, which may not be the case at stagnation points. Thus, flow separation is restricted to corners, cusps and stagnation points. The rigidity condition also implies that

$$\frac{\partial u_k}{\partial t}\frac{\partial A}{\partial x_k}+u_l u_k\frac{\partial^2 A}{\partial x_l \partial x_k}=0, \tag{3.89}$$

where the partial derivative with respect to time is Lagrangian, and all fields are evaluated at a point $x_i=X_i(a_j, s|t)$ on the boundary. It follows that if the boundary is curved, then fluid particles flowing tangentially to the boundary are accelerating normally. These accelerations must be sustained by the normal pressure gradient, or by an externally imposed normal force.

3.9.2 Comoving boundaries

Let S be a surface in label space. It is the locus of labels a_j for which $S(a_j)=0$. The labels may be explicitly expressed as

$$a_j=\zeta_j(b_1, b_2), \tag{3.90}$$

where b_1, b_2 are two surface coordinates. That is,

$$S\big(\zeta_j(b_1, b_2)\big) = 0\,. \tag{3.91}$$

Consider the fluid particles assigned labels on S at time s. At time t, their positions will be given by

$$x_i = \xi_i^t(b_1, b_2) \equiv X_i\big(\zeta_j(b_1, b_2), s|t\big)\,. \tag{3.92}$$

Their locus will be a surface A^t for which there is a function $A^t = A^t(x_j)$ such that

$$A^t\big(\xi_j^t(b_1, b_2)\big) = 0\,. \tag{3.93}$$

Indeed, it is the case that

$$S\Big(A_j\big(\xi_k^t(b_1, b_2), t|s\big)\Big) = S\big(\zeta_j(b_1, b_2)\big) = 0\,, \tag{3.94}$$

since $A_j\big(\xi_k^t(b_1, b_2), t|s\big) = A_j\Big(X_k\big(\zeta_j(b_1, b_2), s|t\big), t|s\Big) = \zeta_j(b_1, b_2)$. Recall the position-to-label transformation $a_i = A_i(x_j, t|s)$, and see Figure 3.3.

The fluid boundary A^t is the map of the label boundary S. The selection of boundary conditions which must be imposed on such a *comoving* boundary will depend upon the dynamics of the fluid.

3.9.3 Comoving boundary conditions

The mass conservation equation (3.3) may be expressed as

$$\frac{\partial \rho}{\partial t} J_s^t + \rho \frac{\partial J_s^t}{\partial t} = 0\,. \tag{3.95}$$

Differentiating (3.95) with respect to time yields

$$\frac{\partial^2 \rho}{\partial t^2} = \frac{2}{\rho}\left(\frac{\partial \rho}{\partial t}\right)^2 - \frac{\rho}{J_s^t}\frac{\partial^2 J_s^t}{\partial t^2}\,. \tag{3.96}$$

In two dimensions, for simplicity,

$$\frac{\partial^2 J_s^t}{\partial t^2} = \frac{\partial^3 X}{\partial^2 t \partial a}\frac{\partial Y}{\partial b} + \frac{\partial X}{\partial a}\frac{\partial^3 Y}{\partial^2 t \partial b} - \frac{\partial^3 X}{\partial^2 t \partial b}\frac{\partial Y}{\partial a} - \frac{\partial X}{\partial b}\frac{\partial^3 Y}{\partial^2 t \partial a} + \cdots \tag{3.97}$$

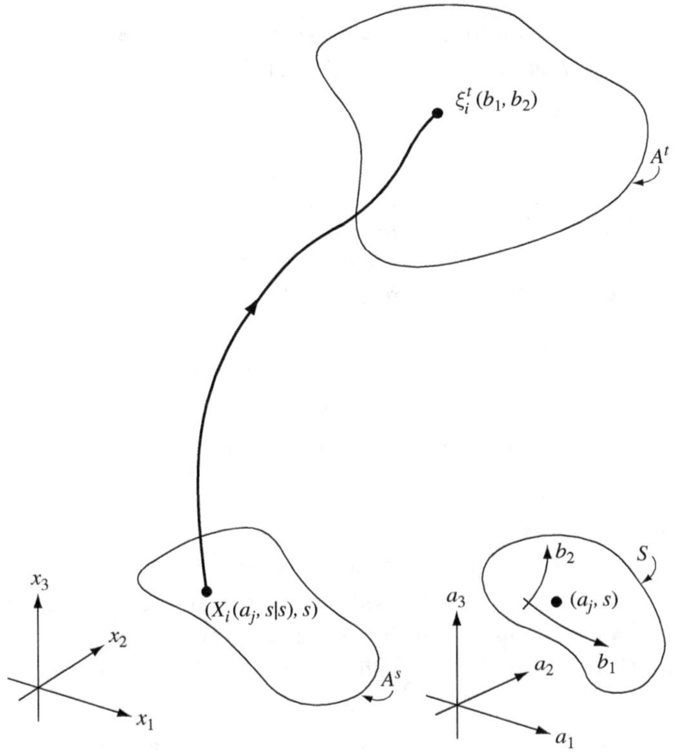

Figure 3.3 At time t, the fluid surface A^t consists of points $\xi_i^t(b_1, b_2)$, where (b_1, b_2) are surface coordinates.

where the ellipsis denotes first-order time and space derivatives of X and Y. The second-order time derivatives are given by the two-dimensional forms of the conservation equations for momentum (3.11), etc.

$$\rho \frac{\partial^2 X}{\partial t^2} = -(J_s^t)^{-1} \left(\frac{\partial p}{\partial a} \frac{\partial Y}{\partial b} - \frac{\partial p}{\partial b} \frac{\partial Y}{\partial a} \right), \tag{3.98}$$

$$\rho \frac{\partial^2 Y}{\partial t^2} = -(J_s^t)^{-1} \left(\frac{\partial X}{\partial a} \frac{\partial p}{\partial b} - \frac{\partial X}{\partial b} \frac{\partial p}{\partial a} \right). \tag{3.99}$$

Note that ρJ_s^t is conserved and so is not a dependent *variable*. Thus, pressure gradients aside, the partial differential equations (3.98) and (3.99) are linear in X and Y. It remains to express the pressure p in terms of the density ρ, and hence in terms of X and Y through J_s^t. Assuming an ideal gas ($p = R\rho T$) having constant heat capacity at constant density ($\mathfrak{E} = C_\rho T$), and assuming homentropic motion (uniform and constant entropy: $\eta \equiv \eta_0$), it follows from the entropy conservation equation (3.15) that

$$p = \theta_0 \rho^\gamma , \tag{3.100}$$

where $\theta_0 = (p_0/\rho_0{}^\gamma)$, while p_0, ρ_0 are a constant pressure and a constant density, respectively, $\gamma = C_p/C_\rho$ being the constant ratio of specific heats. Thus, (3.98) and (3.99) become, respectively

$$\frac{\partial^2 X}{\partial t^2} = -\frac{c^2}{\rho}\frac{1}{J_s^t}\left(\frac{\partial \rho}{\partial a}\frac{\partial Y}{\partial b} - \frac{\partial \rho}{\partial b}\frac{\partial Y}{\partial a}\right) + \cdots \tag{3.101}$$

$$\frac{\partial^2 Y}{\partial t^2} = -\frac{c^2}{\rho}\frac{1}{J_s^t}\left(\frac{\partial X}{\partial a}\frac{\partial \rho}{\partial b} - \frac{\partial X}{\partial b}\frac{\partial \rho}{\partial a}\right) + \cdots \tag{3.102}$$

where $c^2 = \gamma p/\rho$ is the squared sound speed for an ideal gas ($\gamma = 2$ for this two-dimensional monatomic gas with no internal degrees of freedom), and the ellipsis denotes terms in undifferentiated density ρ. Combining (3.96) and (3.97) with (3.101) and (3.102) yields

$$\begin{aligned}\frac{\partial^2 \rho}{\partial t^2} = \left(\frac{c}{J_s^t}\right)^2 \Bigg\{ & \left[\left(\frac{\partial X}{\partial b}\right)^2 + \left(\frac{\partial Y}{\partial b}\right)^2\right]\frac{\partial^2 \rho}{\partial a^2} - 2\left[\frac{\partial X}{\partial a}\frac{\partial X}{\partial b} + \frac{\partial Y}{\partial a}\frac{\partial Y}{\partial b}\right]\frac{\partial^2 \rho}{\partial a \partial b} \\ & + \left[\left(\frac{\partial X}{\partial a}\right)^2 + \left(\frac{\partial Y}{\partial a}\right)^2\right]\frac{\partial^2 \rho}{\partial b^2}\Bigg\} + \cdots \end{aligned} \tag{3.103}$$

where the ellipsis denotes first derivatives of density. Equation (3.103) is hyperbolic, provided J_s^t does not vanish. That possibility is the fundamental pathology for Lagrangian gas dynamics and so will not be discussed at this juncture. The initial density ρ_s is assumed known since it is required for integration of the mass conservation law. The initial time rate of change of density $(\partial \rho/\partial t)_s$ can be inferred from the initial data for the momentum conservation laws, that is, from the particle label (a, b) and initial velocity (u_s, v_s). Values of density ρ suffice as data on a Lagrangian or comoving boundary. The density ρ having been determined everywhere from (3.103), and hence the pressure p from (3.100), it remains to determine the velocity everywhere, and in particular that of the comoving boundary.

The momentum equations (3.98) and (3.99) are partial differential equations for the particle paths, given the pressure field. As will be seen in Section 8.3, they constitute a classically ill-posed problem for the paths, given the pressure field. Velocities must be inferred otherwise.

First rearrange the two-dimensional form of (3.3), the equation of conservation of mass:

$$\frac{\partial J_s^t}{\partial t} = \frac{\partial u}{\partial a}\frac{\partial Y}{\partial b} + \frac{\partial X}{\partial a}\frac{\partial v}{\partial b} - \frac{\partial u}{\partial b}\frac{\partial Y}{\partial a} + \frac{\partial X}{\partial b}\frac{\partial v}{\partial a} = \zeta \tag{3.104}$$

where

$$\zeta = -\frac{J_s^t}{\rho}\frac{\partial \rho}{\partial t}. \tag{3.105}$$

The Cauchy invariant (3.51) is, in two dimensions:

$$\varpi \equiv \frac{\partial X}{\partial b}\frac{\partial u}{\partial a} + \frac{\partial Y}{\partial b}\frac{\partial v}{\partial a} - \frac{\partial X}{\partial a}\frac{\partial u}{\partial b} - \frac{\partial Y}{\partial a}\frac{\partial v}{\partial b} = \varpi_s. \tag{3.106}$$

Eliminating v in favor of u, and conversely, yields

$$\left(M^2+N^2\right)u+\cdots \quad = M\zeta - N\varpi_s \tag{3.107}$$

$$\left(M^2+N^2\right)v+\cdots \quad = N\zeta - M\varpi_s \tag{3.108}$$

where the operators M and N are

$$M \equiv \frac{\partial Y}{\partial b}\frac{\partial}{\partial a} - \frac{\partial Y}{\partial a}\frac{\partial}{\partial b}, \quad N \equiv \frac{\partial X}{\partial a}\frac{\partial}{\partial b} - \frac{\partial X}{\partial b}\frac{\partial}{\partial a} \tag{3.109}$$

and the ellipsis denotes first derivatives of u and v. The right-hand sides of (3.107) and (3.108) are known, since ρ has been determined as a function of time and the initial Cauchy invariant ϖ_s. Considering the elliptic equation (3.107), for example, the Dirichlet boundary condition of prescribed values of u all around determines u everywhere. Finally, both normal *and* tangential derivatives of v follow from (3.104) and (3.106), and so a solution for v compatible with u is determined from (3.108) without further boundary information. Note that it is being argued here that (3.107) and (3.108) may be characterized as though they were linear elliptic equations; indeed, it is being assumed that the path (X, Y) is known at time t, as are the strain components $\partial X/\partial a$, etc. Having solved for (u, v) at time t, the path may then be extrapolated to time $t+\Delta t$, and so on.

In summary, sufficient boundary conditions on a comoving boundary are provided by values of density and normal velocity. These determine all interior fields, and also the subsequent motion of the boundary (Bennett and Chua, 1999).

Incompressible flow may be regarded as the limit of compressible flow at high sound speed c, or more specifically at low Mach number $M_a = (u_i u_i)^{\frac{1}{2}}/c$. In that limit, (3.103) becomes an elliptic equation for density ρ. There is an equivalent elliptic equation for pressure p, and so pressure is determined everywhere by specifying p on comoving boundaries such as waves on water (Kinsman, 1965). The velocity field is determined everywhere, and consequently the subsequent motion of the boundary is also determined, if one component of the velocity is specified on the comoving boundary.

Exercise 3.14 Assume that the flow is not only incompressible, but is also irrotational. Use Bernoulli's theorem (3.62), and Laplace's equation (3.83) to determine the velocity everywhere, given only the pressure on the comoving boundary. The argument is simpler for small-amplitude disturbances. For definiteness, let the force potential be $\mathcal{F}(x_i)=gx_3$, where g is the gravitational constant. Note that the additional assumption of irrotationality obviates the need to specify a velocity component on the comoving boundary. Compare with, for example, Stoker (1957). □

3.9.4 Adjacent Lagrangian coordinates

As demonstrated by Yakubovich and Zenkovich (2001), it can be expedient to use different labeling systems for different sets of fluid particles. The systems must be kinematically and dynamically compatible wherever the systems are adjacent. Let (a_i^1, s^1), (a_i^2, s^2) be two labeling systems. The functional forms for the particle paths labeled by these two systems will in general differ, thus $X_j^1=X_j^1(a_i^1, s^1|t)$, $X_j^2=X_j^2(a_i^2, s^2|t)$. Should the paths be adjacent on the surface $A=A(x_j, t)=0$ at time t, that is, should $A(X_j^1, t)=A(X_j^2, t)=0$, then at least the normal velocity and the pressure must be continuous across the surface:

$$\frac{\partial A}{\partial x_j}\left(\frac{\partial X_j^1}{\partial t}-\frac{\partial X_j^2}{\partial t}\right)=0 \tag{3.110}$$

and

$$p(a_i^1, s^1|t)=p(a_i^2, s^2|t)\,. \tag{3.111}$$

3.10 Local dynamics

The mixed form for conservation of momentum is

$$\frac{\partial^2}{\partial t^2}X_i(a_j, s|t)=-\frac{\partial h}{\partial x_i}[X_k(a_j, s|t), t]\,, \tag{3.112}$$

where the flow is assumed barotropic: $p=p(\rho)$, and h is the enthalpy that satisfies $dh=dp/\rho$. It is now assumed that the label a_k is the particle position at time s. Expanding about a local reference point a_i^* for small $t-s$ yields

$$\frac{\partial^2}{\partial t^2}X_i = -\frac{\partial h}{\partial x_i}[a_j^*, s] - \frac{\partial^2 h}{\partial x_i \partial x_k}[a_j^*, s]\left(X_k - a_k^*\right) + O(|a_j - a_j^*|)^2 + O(t-s). \tag{3.113}$$

Transforming to a new spatial variable ξ_i, according to

$$X_i = a_i^* - \frac{1}{2}(t-s)^2 \frac{\partial h}{\partial x_i}[a_j^*, s] + \xi_i, \tag{3.114}$$

equation (3.113) becomes

$$\frac{\partial^2}{\partial t^2}\xi_i = -\left(\frac{\partial^2 h}{\partial x_i \partial x_j}\right)^* \xi_j + O(|a_j - a_j^*|)^2 + O(t-s). \tag{3.115}$$

The initial position is $\xi_i = a_i - a_i^*$ at $t=s$. Thus the local stability of the particle displacements is determined by the signs of the eigenvalues of the matrix of second partial derivatives of the enthalpy, or Hessian form for the enthalpy. Note that, as for local kinematics, all Eulerian fields are evaluated at $[a_i^*, s]$. The Hessian form is symmetric and so has real eigenvalues which may be positive, zero or negative. If all are zero or positive then the particle orbits are closed. Negative eigenvalues correspond to open orbits: the particle displacements grow exponentially in time. Like the local Lagrangian kinematics analyzed in Section 1.3, local Lagrangian dynamics are necessarily determined by the local Eulerian fields: the rate of strain tensor determines the kinematics, while the Hessian for enthalpy determines the dynamics.

The orders of magnitude of growth rates characterized by (1.33) are consistent with those of (3.115), as a consequence of the local Eulerian dynamics. Specializing to incompressible flow, and taking the divergence of the conservation of momentum yields

$$\frac{\partial^2 p}{\partial x_i \partial x_i} = -\rho \frac{\partial u_i}{\partial x_j}\frac{\partial u_j}{\partial x_i} = -\rho\left(e_{ij}e_{ji} + \sigma_{ij}\sigma_{ji}\right), \tag{3.116}$$

where e_{ij} is a component of the symmetric rate of strain tensor defined by (1.34), while $\sigma_{ij} = -\frac{1}{2}\epsilon_{ijk}\omega_k$ is a component of the skew tensor constructed from the vorticity ω_i: see (1.30). A simplistic analysis of (3.116) would conclude that the order of magnitude scales of the Hessian for the enthalpy p/ρ are the squares of the order of magnitude scales of the rates of strain and the vorticity. For a solenoidal Eulerian velocity field u_i, there is a solenoidal velocity potential χ_i such that

$$u_i = \epsilon_{ijk}\frac{\partial \chi_k}{\partial x_j}. \tag{3.117}$$

Hence the growth rates of particle displacements, according both to local kinematics and to local dynamics, are the order of magnitude scales of the Hessians for the velocity potential. A more complex analysis of (3.116) would recognize that it defines a Poisson problem. The solution should be subject to, for example, homogeneous Neumann conditions at rigid boundaries, and thus the components of the Hessian for enthalpy will depend not only upon the "charge density" on the right-hand side of (3.116), but also upon the geometry of the domain. Even this analysis does not preclude significant differences in the numerical factors associated with the conditioning of the various symmetric matrices. That is, there may be significant differences between the various eigenvalues, and hence between the growth rates according to local kinematics and local dynamics. But in the end the obvious difficulty with local kinematics and local dynamics is of course that the former, but not the latter, only permits closed particle orbits in exceptional cases. This paradox may be understood by recalling that (1.28) and (3.113) are merely the first terms in series expansions of (1.27) and (3.112), respectively. The first term suffices exactly for local kinematics when the Eulerian velocity is a linear function of position, but such a flow is impossible in a rigidly bounded domain. Similarly the first term suffices exactly for local dynamics when the Eulerian pressure is a quadratic function of position, but that too is impossible in a rigid domain. More realistically, it should be conceded that the closed orbits implied by local dynamics are idealizations. Nonuniformity of the Hessian for enthalpy would break up such orbits, and the growth rates of the ensuing secular motions would be consistent with local kinematics. The difficulties encountered when making local analyses emphasize that fluid dynamics is a field theory and, for incompressible flow, is a nonlocal field theory.

3.11 Relabeling symmetry

The unforced motion of an ideal fluid is, according to Exercise 3.4, an extremum of the Lagrange functional

$$L \equiv \int_s^{t_1} \int_{W_a} \Lambda dW_a dt , \tag{3.118}$$

where Λ is the Lagrange density

$$\Lambda = \rho J_s^t \left\{ \frac{1}{2} \frac{\partial X_j}{\partial t} \frac{\partial X_j}{\partial t} - \mathfrak{E}(\rho, \eta) \right\} , \tag{3.119}$$

where W_a and $dW_a = da_1 da_2 da_3$ are respectively the domain and volume element in a_i-labeling space, provided that the extremum is subject to the conditions of conservation of mass:

$$\frac{\partial}{\partial t}\rho J_s^t = 0\,, \tag{3.120}$$

and entropy:

$$\frac{\partial \eta}{\partial t} = 0\,. \tag{3.121}$$

As pointed out in Exercise 3.4, the constraints may be expressed in the forms

$$\rho(a_i, s|t)J(a_i, s|t) = \rho(a_i, s|s)J(a_i, s|s)\,, \tag{3.122}$$

$$\eta(a_i, s|t) = \eta(a_i, s|s)\,. \tag{3.123}$$

It is convenient to make a *time-independent* change of labels from a_i to b_i such that

$$\rho(b_i, s|t)J(b_i, s|t) = \rho(b_i, s|s)J(b_i, s|s) \equiv 1\,. \tag{3.124}$$

Meanwhile,

$$\eta(b_i, s|t) = \eta(b_i, s|s) \equiv \eta_s(b_i)\,, \tag{3.125}$$

say. Then in b_i-label space, the Lagrange functional becomes

$$L = \int_s^{t_1} \int_{W_b} \left(\frac{1}{2}\frac{\partial X_j}{\partial t}\frac{\partial X_j}{\partial t} - \mathfrak{E}\Big(\frac{1}{J_s^t}, \eta_s\Big) \right) dW_b dt\,, \tag{3.126}$$

where the Jacobi determinant is now

$$J_s^t = \frac{\partial(X_1, X_2, X_3)}{\partial(b_1, b_2, b_3)} \tag{3.127}$$

and W_b is the domain in b_i space, with measure element $dW_b = db_1 db_2 db_3$.

There are two cases to consider.

(i) *Homentropic flow*. In this case the entropy η is not only conserved following the motion, its initial and hence subsequent values are uniform throughout the fluid: $\eta(b_i, s|t) \equiv \eta_0$, where η_0 is a constant. It follows that the internal energy $\mathfrak{E}$ and hence the pressure p are functions of density alone: $\mathfrak{E} = \mathfrak{E}(\rho)$, $p = p(\rho)$; that is, the flow is *barotropic*. Now consider a transformation of the label b_i:

$$c_i = C_i(b_j)\,. \tag{3.128}$$

As pointed out by, e.g., Bretherton (1970), Ripa (1981), Salmon (1983) and Zakharov and Kuznetsov (1997), the Lagrange functional (3.126) only retains its form identically if the Jacobi determinant of this relabeling equals unity:

$$\frac{\partial(C_1, C_2, C_3)}{\partial(b_1, b_2, b_3)} = 1\,. \tag{3.129}$$

It is assumed that the boundary is unaltered by the transformation; for example, the boundary remains in the far field where it is assumed that there is no flow. Restricting to infinitesimal transformations such as

$$c_i = b_i + \chi_i(b_j) \tag{3.130}$$

where $O(\chi_i\chi_i)$ is negligible, the invariance condition (3.129) is one of solenoidality:

$$\frac{\partial \chi_i}{\partial b_i} = 0\,. \tag{3.131}$$

Thus an infinitesimal relabeling that leaves the Lagrange functional (3.126) invariant is a curl:

$$\chi_i = \epsilon_{ijk} \frac{\partial \theta_k}{\partial b_j}\,, \tag{3.132}$$

for some infinitesimal vector field $\theta_i(b_j)$. Noether's theorem (e.g, Guidry, 1991; Lanczos, 1966; Ryder, 1996; Weinberg, 1995) guarantees the existence of a conserved quantity for each symmetry of the Lagrange functional, that is, for each transformation that leaves the functional invariant.

Exercise 3.15 Show that, in the case of homentropic flow, and as a consequence of the invariance of the Lagrange functional (3.126) with respect to the infinitesimal transformation (3.132), the following conservation law holds:

$$\frac{\partial}{\partial t}\left(\epsilon_{ijk} \frac{\partial^2 X_m}{\partial t \partial b_j} \frac{\partial X_m}{\partial b_k}\right) = 0\,. \tag{3.133}$$

□

The conserved quantity in (3.133) may be expressed as a curl with respect to b_i; it follows that there is a scalar field ξ such that

$$\frac{\partial}{\partial t}\left(\frac{\partial X_m}{\partial t} \frac{\partial X_m}{\partial b_k}\right) = -\frac{\partial \xi}{\partial b_k}\,, \tag{3.134}$$

and since the transformation from the original label a_i to the label b_i is time independent,

$$\frac{\partial}{\partial t}\left(\frac{\partial X_m}{\partial t}\frac{\partial X_m}{\partial a_k}\right)=-\frac{\partial \xi}{\partial a_k}. \tag{3.135}$$

Taking the curl of (3.135) and rearranging, yields the conservation law

$$\frac{\partial}{\partial t}\varpi_i=0, \tag{3.136}$$

where

$$\varpi_i \equiv \epsilon_{ijk}\frac{\partial^2 X_m}{\partial t \partial a_j}\frac{\partial X_m}{\partial a_k} \tag{3.137}$$

is the Cauchy invariant.

(ii) *Isentropic flow*. In this case entropy η is conserved, but is not uniform with respect to the label b_i. That is, $\eta(b_i, s|t)=\eta(b_i, s|s)\equiv \eta_s(b_i)$. The internal energy $\mathfrak{E}$ and hence the pressure p are not functions of density ρ alone, thus the flow is *baroclinic*. The Lagrange functional (3.126) is invariant with respect to infinitesimal transformations (3.130) that satisfy the solenoidality condition (3.131), and that leave η_s unchanged:

$$\chi_j\frac{\partial \eta_s}{\partial b_j}=0, \tag{3.138}$$

such as (Zakharov and Kuznetsov, 1997)

$$\chi_i=\epsilon_{ijk}\frac{\partial \eta_s}{\partial b_j}\frac{\partial \psi}{\partial b_k} \tag{3.139}$$

for any scalar field ψ.

Exercise 3.16 Show that, in the case of isentropic flow, and as a consequence of the invariance of the Lagrange functional (3.126) with respect to the infinitesimal transformation (3.130), the following conservation law holds:

$$\frac{\partial}{\partial t}\left(\frac{\partial \eta}{\partial a_i}\varpi_i\right)=0. \tag{3.140}$$

Hints:

(i) the following identity holds (Jeffreys, 1931, p. 15):

$$\epsilon_{ijk}\frac{\partial(a_1, a_2, a_3)}{\partial(b_1, b_2, b_3)}=\epsilon_{pqr}\frac{\partial a_p}{\partial b_i}\frac{\partial a_q}{\partial b_j}\frac{\partial a_r}{\partial b_k}; \tag{3.141}$$

(ii) entropy η is conserved;
(iii) mass is conserved.

Deduce from (3.140) that the *Ertel potential vorticity* is conserved:

$$\frac{\partial}{\partial t}\left(\frac{\partial \eta}{\partial x_i}\frac{\omega_i}{\rho}\right)=0, \tag{3.142}$$

where the partial derivative with respect to time is Lagrangian, the partial derivative with respect of position is Eulerian, and ω_i is the Eulerian vorticity. □

Another Lagrange density, discovered by Bateman (1929; 1944, p. 165), also leads to the conservation of momentum as a variational principle. Conservation of mass and conservation of entropy need not be imposed as constraints; rather, they too arise as extremal conditions for the Lagrange functional. Bateman's formulation (see also Eckart, 1960; Seliger and Witham, 1968) is Eulerian, and his action is simply the fluid dynamical presssure p, related to the enthalpy h and entropy η through the combined first and second laws of thermodynamics. The enthalpy is in turn related by Bateman to the potentials in the Clebsch representation of the Eulerian velocity field (Lamb, 1932). The extremal conditions, arising from variations of the Clebsch potentials associated with rotationality of the flow, show that those potentials are conserved following the flow.

The Lagrangian formulation of Bateman's functional is

$$L=\int_s^{t_1}\int_{W_a} J_s^t p dW_a dt. \tag{3.143}$$

The combined first and second laws are

$$Td\eta = dh - \rho^{-1}dp, \tag{3.144}$$

where T is the absolute temperature and ρ is the fluid density.

Exercise 3.17 Introducing potentials ϕ and γ such that

$$h=\frac{1}{2}\frac{\partial X_i}{\partial t}\frac{\partial X_i}{\partial t}+\frac{\partial \phi}{\partial t}+\eta\frac{\partial \gamma}{\partial t}, \tag{3.145}$$

and by considering variations of X_i, ϕ and γ, derive the conservation of momentum, mass and entropy, respectively. By varying η, derive

$$\frac{\partial \gamma}{\partial t}=T. \tag{3.146}$$

Show that the relabeling invariance of the Lagrange functional L yields (3.140), implying the conservation of Ertel's potential vorticity (3.142). Note that the Lagrangian development of Bateman's Lagrange density does not require the introduction of the Clebsch potentials associated with Eulerian vorticity. □

So the powerful constraints of conservation of the Cauchy invariant in barotropic flow, and conservation the Ertel potential vorticity in baroclinic flow, are consequences of the relabeling invariance of the Lagrange functionals. It is widely commented that the derivations of these conservation laws as instances of Noether's theorem are natural, while their derivation from the equations of motion are artificial. Readers of this section may wish to form their own judgements.

3.12 Historical note

Landau and Lifschitz (1959) remark (on their p. 3) that the equation (3.12) here was first obtained by Euler in 1755. Their subsequent treatment (on their p. 5) of the Lagrangian formulation of fluid dynamics is entirely as follows.

PROBLEM

Write down the equations for a one-dimensional motion of an ideal fluid in terms of the variables a, t, where a (called a *Lagrangian variable*†) is the x coordinate of a fluid particle at some instant t_0.

†Although such variables are usually called Lagrangian, it should be mentioned that the equations of motion in these coordinates were first obtained by EULER, at the same time as equations (3.12 here)

Truesdell (1954, p. 30) prefers to call our arbitary position x_i the *spatial coordinate*, and our particle label a_j and time t the *material variables*. He considers as labels only the particle positions at some labeling time s. In particular he "eschews the general misnomer" in which a_j is called a "Lagrangian" coordinate and x_i an "Eulerian coordinate". In an extended footnote, Truesdell recounts the early history of the development of fluid dynamics. Euler "gave a detailed summary of the whole theory of perfect fluids expressed in material variables" in a "letter, date 1 January 1760", to Lagrange. It is evident from Euler's other writings that he was already using the material description some years earlier. For the full account and references, see Truesdell (1954). The "erroneous terminology" used here was introduced by Dirichlet in 1860. Its usage is so prevalent that it cannot be casually abandoned; our alternative title "Material Variables Fluid Dynamics" would not be so widely recognized.

4
Coordinates

4.1 Independent variables

The first Cartesian component of the Lagrangian momentum conservation law is, in familiar notation,

$$\rho \frac{\partial^2 X}{\partial t^2} = -J_a^{-1} \frac{\partial(p, Y, Z)}{\partial(a, b, c)} \tag{4.1}$$

where it is for the moment useful to indicate the particular choice of labels with a representative subscript to the Jacobi determinant:

$$J_a = \frac{\partial(X, Y, Z)}{\partial(a, b, c)} . \tag{4.2}$$

It is important to realize that the fluid particle identifiers or labeling variables (a, b, c) need not be the initial position of the particle. They may for example be values of thermodynamics state variables such as pressure, temperature and salinity at the labeling time. While the need to compute derivatives imposes the requirement that a distance be defined for the labels, the distance need not be Euclidean. That is, the coordinates for the labels need not be Cartesian. Indeed, let (α, β, γ) be any other Lagrangian or labeling variables. The typical momentum equation (4.1) and Jacobi determinant (4.2) transform respectively to

$$\rho \frac{\partial^2 X}{\partial t^2} = -J_\alpha^{-1} \frac{\partial(p, Y, Z)}{\partial(\alpha, \beta, \gamma)} , \tag{4.3}$$

$$J_\alpha = \frac{\partial(X, Y, Z)}{\partial(\alpha, \beta, \gamma)} , \tag{4.4}$$

while conservation of mass

$$\frac{\partial}{\partial t} \left\{ \rho \frac{\partial(X, Y, X)}{\partial(a, b, c)} \right\} = 0 \tag{4.5}$$

transforms to

$$\frac{\partial}{\partial t}\left\{\rho\frac{\partial(X, Y, X)}{\partial(\alpha, \beta, \gamma)}\right\}=0 \tag{4.6}$$

since $\partial(a, b, c)/\partial(\alpha, \beta, \gamma)$ is independent of time t. Conservation laws for scalars, such as (3.15) for entropy, are unchanged. This invariance of form is an alternative manifestation of the relabeling gauge symmetry: see Chapter 3. Finally, note that the labels (a, b, c) and (α, β, γ) need not be assigned at the same time.

4.2 Dependent space variables

The dependent variables in Lagrangian fluid dynamics consist of the fluid particle position, which may be expressed in Cartesian coordinates as $(x, y, z)=(X, Y, Z)$ where $x=X(a, b, c, s|t)$, etc., and a selection of thermodynamic variables such as the density $\rho=\rho(a, b, c, |t)$ and pressure $p=p(a, b, c, s|t)$. Consider transforming the spatial coordinates from Cartesian coordinates (x, y, z) to geographers' spherical polar coordinates (r, ϕ, θ), where r is radial distance, ϕ is longitude and θ is latitude. The defining relationships between the spatial coordinates are

$$x=r\cos\theta\cos\phi,\quad y=r\cos\theta\sin\phi,\quad z=r\sin\theta\,. \tag{4.7}$$

The position of a fluid particle transforms from (X, Y, Z) to (R, Φ, Θ). It is convenient here to use boldface notation for vectors, for example $\mathbf{X}=(X, Y, Z)$. The velocity of a fluid particle is

$$\frac{\partial \mathbf{X}}{\partial t}=\mathbf{u}=u_r\hat{\mathbf{e}}_r+u_\phi\hat{\mathbf{e}}_\phi+u_\theta\hat{\mathbf{e}}_\theta \tag{4.8}$$

where $(\hat{\mathbf{e}}_r, \hat{\mathbf{e}}_\phi, \hat{\mathbf{e}}_\theta)$ is the orthonormal basis, and

$$u_r=\frac{\partial R}{\partial t},\quad u_\phi=R\cos\Theta\frac{\partial\Phi}{\partial t},\quad u_\theta=R\frac{\partial\Theta}{\partial t}\,. \tag{4.9}$$

The partial derivatives with respect to time in (4.8) and (4.9) are Lagrangian. The basis vectors vary with space and time; it is easily shown that their Lagrangian rates of change in time are

$$\frac{\partial\hat{\mathbf{e}}_r}{\partial t}=\quad R^{-1}u_\phi\hat{\mathbf{e}}_\phi+R^{-1}u_\theta\hat{\mathbf{e}}_\theta, \tag{4.10}$$

$$\frac{\partial\hat{\mathbf{e}}_\phi}{\partial t}=-R^{-1}u_\phi\hat{\mathbf{e}}_r+R^{-1}\tan\Theta u_\phi\hat{\mathbf{e}}_\theta, \tag{4.11}$$

$$\frac{\partial\hat{\mathbf{e}}_\theta}{\partial t}=-R^{-1}u_\theta\hat{\mathbf{e}}_r-R^{-1}\tan\Theta u_\phi\hat{\mathbf{e}}_\phi\,. \tag{4.12}$$

The mixed form of the momentum equation is, in Cartesian coordinates,

$$\frac{\partial^2 \mathbf{X}}{\partial t^2} = -\rho^{-1}\nabla p \tag{4.13}$$

where the time derivatives are Lagrangian, but the gradient operator is Eulerian. The particle labels are assumed to be position, but need not be. The mixed form with dependent and independent variables in spherical polar coordinates is, component by component:

$$\frac{\partial^2 R}{\partial t^2} - R\left(\cos\Theta\frac{\partial \Phi}{\partial t}\right)^2 - R\left(\frac{\partial \Theta}{\partial t}\right)^2 = -\rho^{-1}\frac{\partial p}{\partial r}, \tag{4.14}$$

$$R\cos\Theta\frac{\partial^2 \Phi}{\partial t^2} + 2\cos\Theta\frac{\partial R}{\partial t}\frac{\partial \Phi}{\partial t} - 2R\sin\Theta\frac{\partial \Theta}{\partial t}\frac{\partial \Phi}{\partial t} = -(\rho R\cos\Theta)^{-1}\frac{\partial p}{\partial \phi}, \tag{4.15}$$

$$R\frac{\partial^2 \Theta}{\partial t^2} + 2\frac{\partial R}{\partial t}\frac{\partial \Theta}{\partial t} + R\sin\Theta\cos\Theta\left(\frac{\partial \Phi}{\partial t}\right)^2 = -(\rho R)^{-1}\frac{\partial p}{\partial \theta}. \tag{4.16}$$

The Eulerian pressure gradient in (4.14)–(4.16) is readily transformed into spherical polar Lagrangian coordinates (σ, ψ, χ), by expressing the pressure gradient with Jacobi determinants as in (4.1). Meanwhile, mass conservation becomes

$$\begin{aligned}\frac{\partial}{\partial t}\left(\rho\frac{\partial(X,Y,Z)}{\partial(\sigma,\psi,\chi)}\right) &= \frac{\partial}{\partial t}\left(\rho\frac{\partial(X,Y,Z)}{\partial(R,\Phi,\Theta)}\frac{\partial(R,\Phi,\Theta)}{\partial(\sigma,\psi,\chi)}\right)\\ &= \frac{\partial}{\partial t}\left(\rho R^2\cos\Theta\frac{\partial(R,\Phi,\Theta)}{\partial(\sigma,\psi,\chi)}\right) = 0.\end{aligned} \tag{4.17}$$

The striking feature of (4.14)–(4.16) is that they may be expressed in terms of the velocities (4.9) as

$$\frac{\partial}{\partial t}\begin{bmatrix} u_r \\ u_\phi \\ u_\theta \end{bmatrix} + R^{-1}\begin{bmatrix} 0 & -u_\phi & -u_\theta \\ u_\phi & 0 & -\tan\Theta u_\phi \\ u_\theta & \tan\Theta u_\phi & 0 \end{bmatrix}\begin{bmatrix} u_r \\ u_\phi \\ u_\theta \end{bmatrix} = -(\rho R)^{-1}\begin{bmatrix} R\frac{\partial p}{\partial r} \\ (\cos\Theta)^{-1}\frac{\partial p}{\partial \phi} \\ \frac{\partial p}{\partial \phi} \end{bmatrix}. \tag{4.18}$$

The matrix appearing in (4.18) is skew, thus the momentum equation is of the form

$$\mathsf{D}_t\mathbf{u} = -\rho^{-1}\nabla p \tag{4.19}$$

where D_t is a covariant derivative with an $\mathfrak{so}(3)$-valued connection. That is, the derivative involves a connection in the Lie algebra of generators of the special orthonormal group $SO(3)$, or group of three-dimensional rotations having determinant of positive unity: see the next section. However, the relationships (4.9) only involve the simple Lagrangian partial derivative $\partial/\partial t$, rather than the covariant derivative D_t. Thus, as will be seen further in the next section, the system (4.9), (4.19) lacks local invariance with respect to $SO(3)$.

4.3 Rotational symmetry

4.3.1 Globally uniform rotations

A rotation is characterized by a 3×3 orthonormal matrix R:

$$\mathsf{R}\mathsf{R}^T = \mathsf{R}^T\mathsf{R} = \mathsf{I}. \tag{4.20}$$

Such matrices constitute the orthonormal group $O(3)$. Consider a rotation without reflection: $\det(\mathsf{R}) = 1$, that is, consider a matrix R in the special orthonormal group $\mathfrak{so}(3)$. A rotation of the Cartesian particle position is a transformation

$$\mathbf{X} \to \mathbf{X}' = \mathsf{R}\mathbf{X}. \tag{4.21}$$

Now the Jacobi determinant of the particle paths may be expressed as a scalar triple product:

$$J_s^t = \frac{\partial(X, Y, Z)}{\partial(a, b, c)} = \frac{\partial \mathbf{X}}{\partial a} \cdot \frac{\partial \mathbf{X}}{\partial b} \times \frac{\partial \mathbf{X}}{\partial c}, \tag{4.22}$$

and this is clearly invariant under rotations of $\mathbf{X}$ that are independent of position. Equally, the kinetic energy $\rho|\partial \mathbf{X}/\partial t|^2$ is invariant if the rotation is steady. Thus the Lagrange density (3.28) is invariant with respect to global rotations, that is, rotations independent of position or time:

$$\Lambda(\mathbf{X}') \equiv \Lambda' = \Lambda \equiv \Lambda(\mathbf{X}). \tag{4.23}$$

It follows that the vector law of conservation of momentum (3.29) is invariant with respect to global rotations. The scalar laws (3.3), (3.15) and (3.21) for conservation of mass, entropy and energy, respectively, are automatically globally invariant.

4.3.2 Time-varying rotations

Apply a time-varying rotation to the particle position:

$$\mathbf{X}'(\mathbf{a}, s|t) = \mathsf{R}(t)\mathbf{X}(\mathbf{a}, s|t), \tag{4.24}$$

where $\mathbf{a}=(a, b, c)$. The internal energy $\mathfrak{E}=\mathfrak{E}(J^t_s)$ in the Lagrange density Λ is unchanged: $\mathfrak{E}=\mathfrak{E}'$, but the kinetic energy is changed:

$$\rho\left|\frac{\partial \mathbf{X}}{\partial t}\right|^2=\rho\left|\frac{\partial \mathbf{X}'}{\partial t}+\mathsf{A}\mathbf{X}'\right|^2, \tag{4.25}$$

where the *connection* is

$$\mathsf{A}=\mathsf{R}\frac{\partial \mathsf{R}^T}{\partial t}. \tag{4.26}$$

A matrix in $SO(3)$ such as R has only three parameters; for example, three angles $(\theta_1, \theta_2, \theta_3)$:

$$\mathsf{R}=\exp\left(\theta_j \mathsf{K}_j\right) \tag{4.27}$$

where the 3×3 matrices K_j generate $SO(3)$, that is

$$\mathsf{K}_j\equiv\left.\frac{\partial \mathsf{R}}{\partial \theta_j}\right|_{\boldsymbol{\theta}=\mathbf{0}}. \tag{4.28}$$

It is readily seen from (4.20) and (4.26) that the connection A and hence the generators K_j are skew. The latter also satisfy the commutator relations

$$[\mathsf{K}_i, \mathsf{K}_j]=-\epsilon_{ijk}\mathsf{K}_k. \tag{4.29}$$

In fact, $(\mathsf{K}_i)_{jk}=\epsilon_{ijk}$. These generators constitute the basis for the Lie algebra $\mathfrak{so}(3)$, and are independent of space or time; it is the parameters in (4.27) which vary: $\theta_i=\theta_i(t)$. Indeed,

$$\mathsf{A}=\mathsf{R}\frac{\partial \mathsf{R}^T}{\partial t}=-\frac{\partial \theta_k}{\partial t}\mathsf{K}_k. \tag{4.30}$$

The extremal condition for the transformed Lagrange density Λ' becomes

$$\rho J^{t\,\prime}_s\left(\frac{\partial^2\mathbf{X}'}{\partial t^2}+2\mathsf{A}\frac{\partial \mathbf{X}'}{\partial t}+\frac{\partial \mathsf{A}}{\partial t}\mathbf{X}'+\mathsf{A}\mathsf{A}\mathbf{X}'\right)_j=-\frac{\partial p}{\partial a_k}\frac{\partial J^t_s}{\partial\left(\dfrac{\partial X'_j}{\partial a_k}\right)}. \tag{4.31}$$

The skew matrix A may be explicitly expressed in terms of the angle parameters:

$$\mathsf{A}=\begin{bmatrix} 0 & -\frac{\partial\theta_3}{\partial t} & \frac{\partial\theta_2}{\partial t} \\ \frac{\partial\theta_3}{\partial t} & 0 & -\frac{\partial\theta_1}{\partial t} \\ -\frac{\partial\theta_2}{\partial t} & \frac{\partial\theta_1}{\partial t} & 0 \end{bmatrix}. \tag{4.32}$$

Hence, being left-multiplied by the matrix A is equivalent to being left-multiplied in a vector product:

$$\mathsf{A}\mathbf{X}'=\boldsymbol{\Omega}\times\mathbf{X}', \tag{4.33}$$

where $\boldsymbol{\Omega} = \partial \boldsymbol{\theta} / \partial t$, and so (4.31) is equivalent to (Sattinger and Weaver, 1986):

$$\rho J_s^{t\,\prime} \left(\frac{\partial^2 \mathbf{X}'}{\partial t^2} + 2\boldsymbol{\Omega} \times \frac{\partial \mathbf{X}'}{\partial t} + \frac{\partial \boldsymbol{\Omega}}{\partial t} \times \mathbf{X}' + \boldsymbol{\Omega} \times \boldsymbol{\Omega} \times \mathbf{X}' \right)_j = -\frac{\partial p}{\partial a_k} \frac{\partial J_s'}{\partial \left(\frac{\partial X_j'}{\partial a_k} \right)} . \tag{4.34}$$

Equation (4.34) may be recognized as the Lagrangian form of conservation of momentum in an unsteadily rotating reference frame.

Spatially nonuniformly rotations may be considered: $\mathsf{R} = \mathsf{R}(\mathbf{a}, s|t)$. These lead to even greater complexity in the transformed Jacobi determinant in Λ, and hence in the pressure gradient in the momentum equation. That is, the form of the momentum equation is not invariant with respect to rotations which are functions of time or position: the momentum equation lacks *local* $SO(3)$ invariance.

Exercise 4.1 Extend the conservation laws (3.136) and (3.140) to a rotating reference frame. □

5
Real fluids

5.1 Viscous stresses and heat conduction

It has been assumed to this point that there are no viscous stresses, nor any heat conduction. Thus, the dynamics of the ideal fluid of the preceding chapters are compatible with an isotropic distribution of molecular velocities. In fact, anisotropy is always present in a real assembly of molecules, owing to the walls of the fluid container, fields of force or sources of heat. The Navier–Stokes equations for a real fluid may be derived from Boltzmann's equation for a dilute gas using the Chapman–Enksog expansion (Chapman and Cowling, 1970), which assumes a molecular velocity distribution close to an isotropic equilibirum. A simpler derivation, requiring less physical insight, follows from the general principles of continuum mechanics by adopting Newton's and Fourier's laws as the constitutive relations. The essential aspect of these constitutive relations is that they are local in the Eulerian framework: the viscous stress tensor is proportional to the Eulerian rate of strain tensor, while the heat flux is proportional to the Eulerian temperature gradient. The Navier–Stokes equations are accordingly expressed naturally in Eulerian form, while the Lagrangian form can only be derived by "cheating." That is, it cannot be derived from Boltzmann's equation. Cheating can be minimized (see *Aside* in Section 3.2), but in the interest of moving forward, let us cheat in full.

5.2 Navier–Stokes equations for incompressible flow

The Eulerian forms of the Navier–Stokes equations for incompressible flow are

$$\frac{\partial u_i}{\partial t} + u_j \frac{\partial u_i}{\partial x_j} = -\rho^{-1} \frac{\partial p}{\partial x_i} + \nu \frac{\partial^2 u_i}{\partial x_k \partial x_k}, \tag{5.1}$$

$$\frac{\partial u_j}{\partial x_j} = 0, \tag{5.2}$$

where ρ is the constant density, ν is the assumed constant kinematic viscosity for shear and x_k is a Cartesian coordinate. In order to transform the Laplacian operator from Eulerian to Lagrangian variables, it suffices to note that, in two dimensions for simplicity,

$$\frac{\partial f}{\partial x} = \frac{\partial(f, Y)}{\partial(x, y)} = (J_s^t)^{-1} \frac{\partial(f, Y)}{\partial(a, b)} \tag{5.3}$$

where f is any function. The Jacobi determinant J_s^t is independent of time for incompressible flow; for simplicity assume that $J_s^t \equiv 1$, as is the case if the labels are the initial positions of the fluid particles. The Laplacian operator in (5.1) may be transformed by repeated applications of the rule (5.3). The Lagrangian form of the planar momentum equations (5.1) then become

$$\frac{\partial^2 X}{\partial t^2} = -\rho^{-1} \frac{\partial(p, Y)}{\partial(a, b)} + \nu \left[\left(\frac{\partial Y}{\partial b}\frac{\partial}{\partial a} - \frac{\partial Y}{\partial a}\frac{\partial}{\partial b} \right)^2 + \left(\frac{\partial X}{\partial a}\frac{\partial}{\partial b} - \frac{\partial X}{\partial b}\frac{\partial}{\partial a} \right)^2 \right] \frac{\partial X}{\partial t} \tag{5.4}$$

and

$$\frac{\partial^2 Y}{\partial t^2} = -\rho^{-1} \frac{\partial(X, p)}{\partial(a, b)} + \nu \left[\left(\frac{\partial Y}{\partial b}\frac{\partial}{\partial a} - \frac{\partial Y}{\partial a}\frac{\partial}{\partial b} \right)^2 + \left(\frac{\partial X}{\partial a}\frac{\partial}{\partial b} - \frac{\partial X}{\partial b}\frac{\partial}{\partial a} \right)^2 \right] \frac{\partial Y}{\partial t}. \tag{5.5}$$

The operator in the square brackets in (5.4) and (5.5) is of second order, and is elliptic since $J_s^t \equiv 1$. Hence these are effectively parabolic equations for the Lagrangian velocities, and the viscosities are effectively unbounded from above. The classical well-posedness of incompressible, viscous flow in two dimensions has been established (Ladyzhenskaya, 1969), but remains an open question in three dimensions (Temam, 2000). Questions of existence of classical solutions aside, we shall assume the equations to be candidates for well-posed initial value problems.

Exercise 5.1 Show that the Lagrangian form of (5.1) is

$$\frac{\partial^2 X_i}{\partial t^2} = -\rho^{-1} \frac{\partial A_j}{\partial x_i}\frac{\partial p}{\partial a_j} + \nu \left(\frac{\partial^2 A_l}{\partial x_k \partial x_k}\frac{\partial u_i}{\partial a_l} + \frac{\partial A_j}{\partial x_k}\frac{\partial A_l}{\partial x_k}\frac{\partial^2 u_i}{\partial a_j \partial a_l} \right). \tag{5.6}$$

The shorthand in (5.6) is, for example, labeling by position,

$$\frac{\partial A_j}{\partial x_i} \equiv \frac{\partial}{\partial x_i} X_j(x_n, t|s) \Big|_{x_n = X_n(a_m, s|t)}. \tag{5.7}$$

□

Exercise 5.2 If the fluid has a constant thermal conductivity k_T, then the energy equation (3.12) becomes

$$T\frac{\partial \eta}{\partial t} = \frac{\partial \mathfrak{E}}{\partial t} + p\frac{\partial \left(\rho^{-1}\right)}{\partial t} = \frac{k_T}{\rho}\frac{\partial^2 T}{\partial x_j \partial x_j}, \tag{5.8}$$

where T is the temperature, and x_j is a Cartesian coordinate. This equation is mixed: the time derivatives are Lagrangian, while the space derivatives are Eulerian. Transform the Laplacian of temperature into Lagrangian variables. Note that (5.8) does not include energy loss owing to viscous dissipation. Should it? Express the loss in Lagrangian form. □

5.3 Matrix formulation for viscous incompressible flow

The matrix formulation of the Lagrangian equations for inviscid compressible flow (see Section 3.7.2) has been extended by Yakubovich and Zenkovich (2001) to viscous incompressible flow. First, assemble the Cauchy invariants ϖ_i into a skew matrix S where

$$S_{ij} = \epsilon_{ijk}\varpi_k, \tag{5.9}$$

which may be disassembled as $\varpi_i = \frac{1}{2}\epsilon_{ijk}S_{jk}$. Then the matrix of invariants is related to the Jacobi matrix by

$$\mathsf{S} = \frac{\partial \mathsf{J}^T}{\partial t}\mathsf{J} - \mathsf{J}^T\frac{\partial \mathsf{J}}{\partial t}. \tag{5.10}$$

Recall the consistency condition for the Jacobi matrix:

$$\frac{\partial J_{ij}}{\partial a_k} = \frac{\partial J_{ik}}{\partial a_j}, \tag{5.11}$$

and recall also the conservation of mass:

$$\frac{\partial}{\partial t}\left(\rho \ \det(J_{ij})\right) = 0. \tag{5.12}$$

The definition (5.10), the identity (5.11) and the mass conservation law (5.12) all hold for viscous compressible flow. However, whereas S is conserved for irrotationally forced inviscid barotropic flow, it obeys a homogeneous diffusion-like equation for irrotationally forced viscous incompressible flow. This equation is most easily expressed in terms of the vector of Cauchy invariants $\boldsymbol{\varpi}$:

$$\frac{\partial \boldsymbol{\varpi}}{\partial t} = -\nu\boldsymbol{\nabla}\times\left(\sqrt{g}^{-1}\mathsf{g}\boldsymbol{\nabla}\times\left(\sqrt{g}^{-1}\mathsf{g}\boldsymbol{\varpi}\right)\right), \tag{5.13}$$

where again ν is the kinematic viscosity, while

$$(\boldsymbol{\nabla} \times \boldsymbol{v})_i = \epsilon_{ijk} \frac{\partial v_k}{\partial a_j}, \tag{5.14}$$

and $\mathsf{g} = \mathsf{J}^T \mathsf{J}$ is the symmetric metric tensor, with determinant g, encountered in Section 3.8. If the labels a_i are the initial positions of the fluid particles, then $g \equiv 1$ for incompressible flow. The vector operator $\boldsymbol{\nabla}$ is the gradient in the space of the label $\mathbf{a}$. It is evident from (5.13) that $\boldsymbol{\varpi}$ is not an invariant in viscous flow.

Exercise 5.3 Derive (5.13).
Hints:

(i) for any smooth vector field $\boldsymbol{v}$,

$$\boldsymbol{\nabla} \cdot \boldsymbol{\nabla} \times \boldsymbol{v} = \mathbf{0}, \tag{5.15}$$

$$\boldsymbol{\nabla} \times \boldsymbol{\nabla} \times \boldsymbol{v} = \boldsymbol{\nabla}\boldsymbol{\nabla} \cdot \boldsymbol{v} - \nabla^2 \boldsymbol{v}; \tag{5.16}$$

(ii) for any Cartesian tensor S with components S_{ij},

$$\epsilon_{ijk} \det(\mathsf{S}) = \epsilon_{lmn} S_{il} S_{jm} S_{kn}. \tag{5.17}$$

See for example, Jeffreys (1931), Chapter 1, Example 7. □

5.4 Boundary conditions

The question arises: what boundary conditions are appropriate for the Lagrangian formulation of viscous, conducting incompressible flow? The state of the flow at the boundary is a matter of physics, and is independent of the mathematical formulation. There is no better way to begin than by quoting Goldstein (1965, p. 677): "The question of the conditions to be satisfied by a moving fluid in contact with a solid body was one of considerable difficulty for a long time, and its importance will justify a short historical note. At the present time it appears to be definitely settled that for practical purposes the fluid immediately in contact with a solid body may be taken as having no velocity relative to the solid, at any rate for nearly all fluids; but the exact conditions on a molecular scale remain still in doubt." This then, is our answer for flow in contact with a solid body at rest: the flow is also at rest. In Lagrangian terms, this implies that a particle released or labeled on a solid boundary will remain at the same position on the boundary for all time, and any particle not in contact with the boundary at the labeling time, or at any other, will never be nor ever was in contact with the boundary.[1] Let A be a fixed, solid boundary defined as the locus of points x_i for which

$$A(x_i)=0. \tag{5.18}$$

Then, as a particle can be on the boundary if and only if it was so located at the labeling time s:

$$A(X_i(a_j, s|t))=0 \quad \Longleftrightarrow \quad A(X_i(a_j, s|s))=0. \tag{5.19}$$

This condition holds even if there is slip around the boundary, and so does not completely characterize viscous flow. The complete characterization of viscous flow is the no-slip condition: a particle initially on the boundary does not move at any time:

$$A(X_i(a_j, s|s))=0 \quad \Longrightarrow \quad X_i(a_j, s|t)=X_i(a_j, s|s). \tag{5.20}$$

Note that the Lagrangian label a_i need not be the position of the particle at time s.

Consider next a comoving boundary within a viscous fluid, that is, a surface consisting of identified fluid particles. The physical conditions which hold on such a surface, or indeed on any surface within the fluid, are those of continuity of normal velocity and continuity of stress. It has already been determined in Section 3.9.3 that normal velocity and density (or equivalently pressure, in isentropic flow) must be specified on a comoving boundary in an ideal compressible fluid. Incompressible, irrotational wave theory calls for only for pressure at free surfaces (see Exercise 3.9.3). For incompressible rotational flow, the comoving boundary values of normal velocity, shear stress components and pressure just determine the solutions of the formally parabolic momentum equations and the elliptic equation for pressure.

Finally, the solution of the energy equation (5.8), the latter naturally expressed as a parabolic equation for temperature, is determined by either the boundary temperature or by the conductive heat flux across the boundary.

[1] The philosophy of the continuum abstraction is now painfully clear. A fluid particle is a spatially small parcel containing a large number of molecules; the parcel does not slip even if some of the molecules do.

PART II

Lagrangian Flows

Introduction

Few analytical solutions are known for the Lagrangian formulation of fluid dynamics; as few, in fact, as are known for the Eulerian. Almost all these solutions describe flow free of momentum advection. The Gerstner waves, and their generalizations the Ptolemaic vortices, stand out as extraordinary exceptions. The classical investigative techniques of linearized hydrodynamic stability theory are available to both the Lagrangian and the Eulerian formulation, as are the newer techniques of phase plane analysis. The classical Stokes' problems for viscous flow near plates are solvable in both formulations; the Lagrangian self-similar solution for Blasius' approximate boundary layer dynamics is intriguingly more complicated than the Eulerian. The general solvability of the Lagrangian formulation of inviscid incompressible fluid dynamics appears, to an applied mathematician, to depend upon the choice of dependent variables. The classical solvability of the viscous problem comes exasperatingly close to being provable in the large, but in the end remains an open question for pure mathematicians.

6

Some analytical Lagrangian solutions

There are few known analytical solutions of the equations of hydrodynamics in Lagrangian form. Then again, few are known for the equations in Eulerian form either. Some elementary analytical Lagrangian solutions, and some failed attempts at such solutions, are presented here. More complex complete or partial Lagrangian solutions may be found in the subsequent two chapters.

6.1 Flow around a cylinder

Consider flow around a rigid cylinder of radius R centered at the origin: $x_i = 0$, and with its axis in the x_3 direction. For familiarity let $(x_1, x_2, x_3) = (x, y, z)$, $(a_1, a_2, a_3) = (a, b, c)$, etc. The surface of the cylinder is defined by

$$A(x, y, z) = x^2 + y^2 - R^2 = 0\,. \tag{6.1}$$

Far from the cylinder, the flow is uniform and steady in the x direction:

$$(u, v, w) \sim (U, 0, 0) \tag{6.2}$$

as $x^2 + y^2 + z^2 \to \infty$. For irrotational flow, there is a velocity potential χ: $u = \partial\chi/\partial x$, etc. If the flow is also incompressible: $\partial u/\partial x + \cdots = 0$, then χ is harmonic: $\partial^2\chi/\partial x^2 + \cdots = 0$. These conditions are more easily exploited in this Eulerian form, and may be solved using images. The flow is axially symmetric, so the axial or third dimension $(c, z, w, \ldots,$ etc.) will no longer be considered. In the interest of preserving the notational environment, the symbols c, z and w will now be recycled. Introducing the complex Eulerian coordinate z and complex Eulerian velocity w, defined as

$$z = x + iy, \quad w = u - iv \tag{6.3}$$

where $i=\sqrt{-1}$, the image solution is well known to be

$$w=U-\frac{UR^2}{z^2}. \tag{6.4}$$

Let Z be the complex particle position:

$$Z=Z(c, s|t) \tag{6.5}$$

where $c=a+ib$ is the complex labeling position at time $t=s$. For steady flow, $w[Z, t]=w[Z]$; indeed, from (6.3),

$$\frac{\partial Z}{\partial t}=\overline{w[Z]}=U-\frac{UR^2}{\overline{Z}^2}, \tag{6.6}$$

where the overbar denotes complex conjugation. It is readily seen from (6.6) that

$$\frac{\partial}{\partial t}(Z\bar{Z})=0 \tag{6.7}$$

when $Z\bar{Z}=R^2$. Thus fluid particles initially on the cylinder remain on it. The nonholomorphic differential equation (6.6) does not have an elementary integral. However, the conditions of incompressibility and irrotationality imply that w is an analytic or holomorphic function of z, and hence there is a complex velocity potential $\Pi=\Pi(z)$ such that $w=d\Pi/dz$. Indeed,

$$\Pi=\chi-i\psi=Uz+\frac{UR^2}{z}, \tag{6.8}$$

where χ and ψ are, respectively, the real velocity potential and streamfunction. Since particle paths coincide with streamlines in steady flow, it follows that if (x, y) is a particle path passing through (a, b) at time s, then ψ is a constant on the path, and is given by

$$\psi=-UY+\frac{UR^2Y}{X^2+Y^2}=-Ub+\frac{UR^2b}{a^2+b^2}. \tag{6.9}$$

This implicit relationship between X and Y tells the value of neither at any time t on the path. An explicit analytical form for the Lagrangian path does not appear available. Simultaneous numerical integration of the real and imaginary parts of (6.6) is straightforward: see Figure 6.1

6.2 Gerstner's trochoidal wave

This classic planar solution (e.g., Lamb, 1932, Art. 251; Milne-Thompson, 1960, Section 14.80; Rankine, 1863) is explicitly derivable only in Lagrangian form. For familiarity let the as yet undefined labels (a, b) be assigned at time $t=s$,

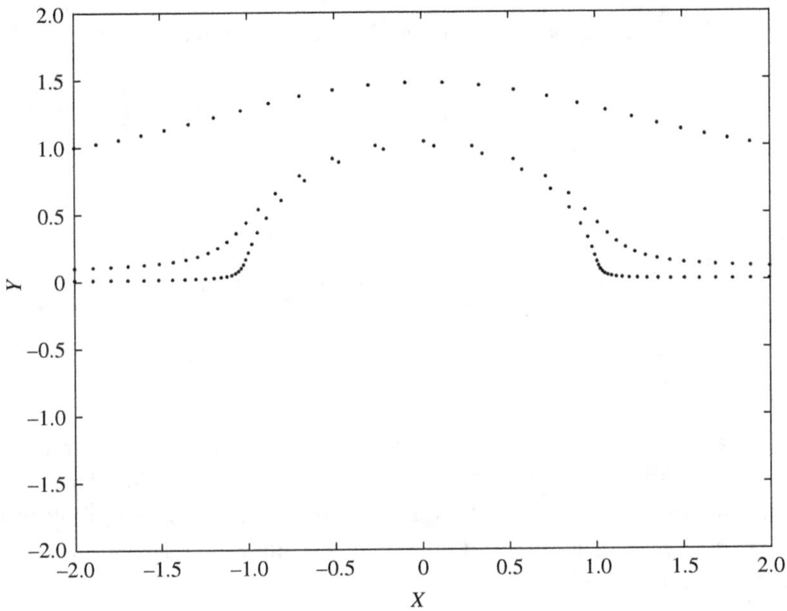

Figure 6.1 Particles flowing over a circular cylinder of radius $R=1$. The far field velocity is $(U,0,0)=(1,0,0)$. The particles are released at $(a,b)=(-2,0.01)$, $(-2,0.1)$ and $(-2,1.0)$, at the constant time interval $\Delta t=10/69$.

with particle paths $X(a,b,s|t)$, $Y(a,b,s|t)$. In the presence of a gravitational acceleration g in the negative y direction, the Lagrangian equations of motion are:

$$\frac{\partial^2 X}{\partial t^2} = -\frac{1}{J_s^t}\frac{\partial(p,Y)}{\partial(a,b)}, \tag{6.10}$$

$$\frac{\partial^2 Y}{\partial t^2} = -\frac{1}{J_s^t}\frac{\partial(X,p)}{\partial(a,b)} - g, \tag{6.11}$$

where a constant density $\rho=1$ has been assumed. The flow is incompressible:

$$\frac{\partial}{\partial t}J_s^t = \frac{\partial}{\partial t}\frac{\partial(X,Y)}{\partial(a,b)} = 0. \tag{6.12}$$

The Gerstner solution is:

$$X = a + \frac{1}{m}\exp(mb)\sin\theta, \tag{6.13}$$

$$Y = b - \frac{1}{m}\exp(mb)\cos\theta \tag{6.14}$$

where $\theta = m\big(a+ct\big)$, while m and c are constants to be determined.

Exercise 6.1 Verify that J_s^t is an invariant, but not equal to unity:

$$J_s^t = 1 - \exp(2mb)\,; \tag{6.15}$$

indeed,

$$X(a,\,b,\,s|s) = a + \frac{1}{m}\exp(mb)\sin\big(m(a+cs)\big) \neq a\,, \tag{6.16}$$

$$Y(a,\,b,\,s|s) = b - \frac{1}{m}\exp(mb)\cos\big(m(a+cs)\big) \neq b\,. \tag{6.17}$$

Thus, the release position is not at $(a,\,b)$, but lies instead on a circle with center $(a,\,b)$ and radius $r = m^{-1}\exp(mb)$. The particle remains on this circle for all time: see Figure 6.2. Note that the release position involves the wavenumber m and the wave amplitude r, but not the phase speed c which is as yet undetermined. □

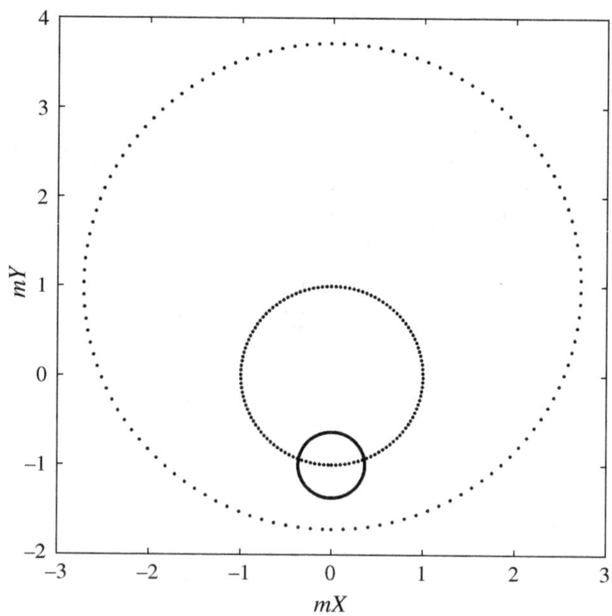

Figure 6.2 The Gerstner trochoidal wave: scaled particle orbits (mX, mY), centered at $(ma, mb) = (0, -1)$, $(0, 0)$ and $(0, 1)$. The phase $\theta = m(a+ct)$ has increments $\triangle\theta = 0.05$.

The pressure p is most easily found after left-multiplying the momentum equations by the Jacobi matrix:

$$\frac{\partial X}{\partial a}\frac{\partial^2 X}{\partial t^2}+\frac{\partial Y}{\partial a}\left(\frac{\partial^2 Y}{\partial t^2}+g\right)=-\frac{\partial p}{\partial a}, \tag{6.18}$$

$$\frac{\partial X}{\partial b}\frac{\partial^2 X}{\partial t^2}+\frac{\partial Y}{\partial b}\left(\frac{\partial^2 Y}{\partial t^2}+g\right)=-\frac{\partial p}{\partial b}. \tag{6.19}$$

Exercise 6.2 Show that the pressure field is

$$p(a,b,s|t)=q(s|t)-gb+\frac{c^2}{2}\exp(2mb)+\frac{(g-mc^2)}{m}\exp(mb)\cos\theta \tag{6.20}$$

where q and c remain arbitrary. It is clear that there can be no surface of constant pressure (to be identified with the ocean surface) unless $q=0$ and $g=mc^2$, in which case p is independent of time for all labels. If the dispersion relation $c^2=g/m$ does hold, then the group velocity in the labeling space is given by

$$c_g=\frac{\partial(mc)}{\partial m}=\frac{c}{2}. \tag{6.21}$$

The waves are dispersive: $c_g\neq c$. □

In the general case $\left(c^2\neq g/m\right)$, the Cauchy–Weber integrals are

$$\frac{\partial X}{\partial a}u+\frac{\partial Y}{\partial a}v=-\frac{\partial\varphi}{\partial a}+\alpha, \tag{6.22}$$

$$\frac{\partial X}{\partial b}u+\frac{\partial Y}{\partial b}v=-\frac{\partial\varphi}{\partial b}+\beta \tag{6.23}$$

where, as is readily verified, $\varphi=-(c/m)\exp(mb)\sin\theta$, $\alpha=c\exp(2mb)$ and $\beta=0$. For φ, see Figure 6.3.

Exercise 6.3 Determine the Cauchy invariant ϖ and the vorticity ω. Consider the sign of $\omega/(mc)$. □

Exercise 6.4 Verify that the particle orbits and velocities in Gerstner's wave satisfy the labeling theorem for arbitrary labels (see Exercise 1.4). □

The Gerstner wave has been extended by Pollard (1970) to a vertically stratifed rotating fluid, and by Constantin (2001) to an edge wave on a uniformly sloping beach. It will be seen in Chapter 7 that the Gerstner wave is one of the *Ptolemaic vortices*.

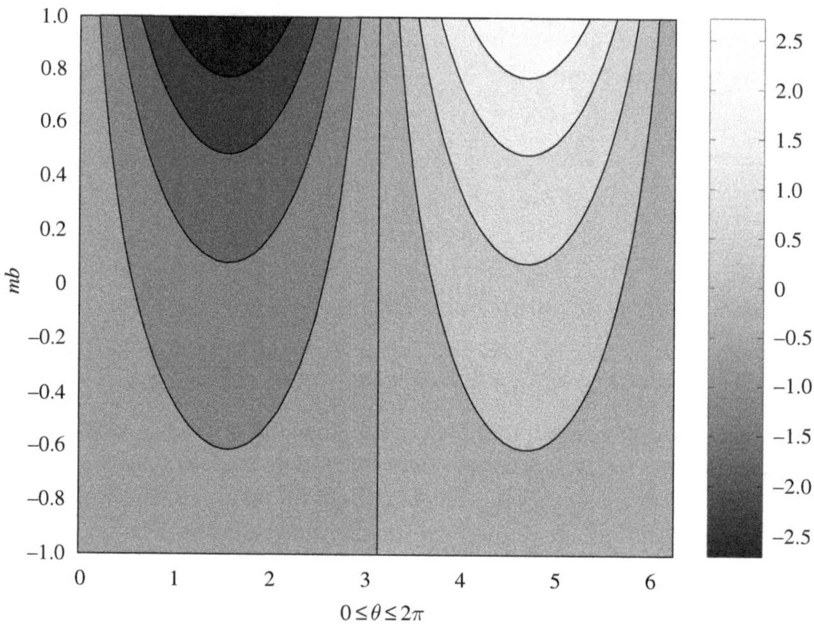

Figure 6.3 The Gerstner trochoidal wave: contours of the scaled Cauchy–Weber integral scalar $m\varphi/c$ as a function of the scaled label mb and the phase θ.

6.3 One-dimensional gas dynamics

6.3.1 One-dimensional traveling waves

Finite-amplitude velocity solutions of the form $u = u(\rho)$, where ρ is the density, may be found for the Eulerian form of the equations of motion of isentropic gas flow (Landau and Lifschitz, 1959, Section 94). Indeed, the general solution is obtained by integrating

$$\left(\frac{\partial x}{\partial t}\right)_u = u + \frac{1}{\rho}\frac{dp}{du} = u \pm c(u) \tag{6.24}$$

where c is the sound speed. Note that $p = p(\rho)$ in homentropic flow, so only one independent variable is needed; in the above equation it is the fluid velocity u. The integral is

$$x = t\big(u \pm c(u)\big) + f(u)\,, \tag{6.25}$$

where f is an arbitrary function of u. The analogous differential equation in Lagrangian form is

$$\left(\frac{\partial a}{\partial t}\right)_u = \pm c(u)\left(\frac{\partial X}{\partial a}\right)^{-1}. \tag{6.26}$$

The strain $\partial X/\partial a$ is not a function of u alone since X is the integral in time of the velocity u, along a particle path which need not be a path of constant u. Thus, $\partial X/\partial a$ depends more generally upon the Lagrangian variables $(a, s|t)$ and so a simple integral is not available.

6.3.2 Riemann invariants

The Riemann invariants for one-dimensional isentropic flow of an ideal gas are the same in Eulerian and Lagrangian form:

$$R_{\pm} = u \pm \left(\frac{2}{\gamma - 1}\right) c, \tag{6.27}$$

where the sound speed c is given by

$$c = \left(\frac{\gamma p}{\rho}\right)^{\frac{1}{2}}, \tag{6.28}$$

and γ is the ratio of specific heats: $\gamma = C_p/C_\rho$. The quantities $R_{\pm}$ are invariant along the Lagrangian characteristics $a = \alpha_{\pm}(t)$, respectively, where

$$\frac{d\alpha_{\pm}}{dt} = \pm c\left(\frac{\partial X}{\partial a}\right)^{-1}. \tag{6.29}$$

6.3.3 Arbitrary one-dimensional flow

Landau and Lifschitz (1959, Section 98) consider arbitrary one-dimensional isentropic gas flow. The Eulerian independent variables x and t are replaced with the fluid velocity u and the enthalpy h. Recall that $d\eta = 0$ in isentropic flow, and so $dh = dp/\rho$ is integrable.

The equations of motion are eventually reduced to a linear partial differential equation in the new variables; families of solutions are known for monatomic and diatomic ideal gases, for example.

Replacing the Lagrangian independent variables a and t in an analogous fashion does not appear to lead to a linear equation; for example, the Lagrangian conservation of mass is

$$\frac{\partial}{\partial t}\left(\rho \frac{\partial X}{\partial a}\right) = 0, \tag{6.30}$$

and so for all time t,

$$\rho(a, s|t)\frac{\partial X}{\partial t}(a, s|t) = \rho(a, s|s) = \rho_s(a) \tag{6.31}$$

where ρ_s, the density at the labeling time, is an arbitrary function of a. Thus, demoting the label a to the rank of dependent variable guarantees the arbitrary nonlinearity of the problem.

6.4 Plane Poiseuille flow

Consider steady, planar incompressible flow with constant kinematic viscosity ν. Assume that the flow is restricted to the x direction and all fields are independent of x, and let us cheat for now. The Eulerian equations of momentum conservation reduce to

$$\frac{\partial^2 u}{\partial y^2} = 0. \tag{6.32}$$

This is readily converted to Lagrangian form, using the identity

$$\frac{\partial u}{\partial y} = \frac{\partial(x, u)}{\partial(x, y)} = \frac{\partial(X, u)}{\partial(a, b)} \equiv J(X, u) \tag{6.33}$$

since the Jacobi determinant of the transformation $(a, b) \to (X, Y)$ is a constant: $J(X, Y) = 1$, say. Thus

$$J\big(X, J(X, u)\big) = 0. \tag{6.34}$$

The condition of longitudinally uniform flow is

$$\frac{\partial u}{\partial x} = J(u, Y) = 0. \tag{6.35}$$

It follows that $u = U(Y)$ for some function U. But then $U''(Y) = 0$, according to (6.34), and so

$$U(Y) = A + BY, \tag{6.36}$$

where A and B are constants. The particle paths originating at (a, b) at time $t = s$ are

$$X = a + (A + Bb)(t - s). \tag{6.37}$$

$$Y = b, \tag{6.38}$$

which does not deserve a figure.

7

Sound waves, shear instabilities, Rossby waves and Ptolemaic vortices

7.1 Sound waves

The equation for finite-amplitude sonic disturbances has been derived in Section 3.9.3. It is derived in this section for infinitesimal sound waves, as a tutorial for the linearized hydrodynamic instability theory developed in the following section.

Again, the Lagrangian form of conservation of mass is

$$\frac{\partial(\rho J_s^t)}{\partial t} = 0. \tag{7.1}$$

After multiplying by the Jacobi matrix, the Lagrangian form of the momentum conservation equation is

$$\rho \frac{\partial X_i}{\partial a_k} \frac{\partial u_i}{\partial t} = -\frac{\partial p}{\partial a_k}, \tag{7.2}$$

while conservation of entropy is

$$T\frac{\partial \eta}{\partial t} = \frac{\partial \mathfrak{E}}{\partial t} + p\frac{\partial(\rho^{-1})}{\partial t} = 0. \tag{7.3}$$

For an ideal calorifically perfect gas,

$$p = R\rho T, \quad \mathfrak{E} = C_\rho T \tag{7.4}$$

where R is the gas constant, and C_ρ is the constant specific heat capacity at constant density. For homentropic motion, it follows that

$$p = p_0 \left(\frac{\rho}{\rho_0}\right)^\gamma \tag{7.5}$$

where the uniform positive constants p_0 and ρ_0 are, respectively, a pressure and a density, while $\gamma = (C_\rho + R)/C_\rho$.

Now suppose that all disturbances p', etc., are small: $p = p_0 + p'$, $|p'| \ll p_0$, for example. Then $dp' = c_0^2 d\rho'$ where c_0 is the constant speed of sound in the static state:

$$c_0^2 = \left(\frac{\partial p}{\partial \rho}\right)_\eta \bigg|_0 = \gamma \frac{p_0}{\rho_0} = \gamma R T_0 \,. \tag{7.6}$$

Assuming that the particle labels are the particle positions a_j at time s, the particle paths may be expressed as

$$X_i(a_j, s|t) = a_i + \xi'_i(a_j, s|t) \,. \tag{7.7}$$

Hence

$$\frac{\partial X_i}{\partial a_j} = \delta_{ij} + \frac{\partial \xi'_i}{\partial a_j} \,, \tag{7.8}$$

and

$$J_s^t = \det\left(\frac{\partial X_i}{\partial a_j}\right) = 1 + \frac{\partial \xi'_k}{\partial a_k} + O(\xi')^2 \,. \tag{7.9}$$

After dropping primes, the equations of motion for infinitesimal disturbances become

$$\frac{\partial \rho}{\partial t} = -\rho_0 \frac{\partial^2 \xi_k}{\partial t \partial a_k} \,, \tag{7.10}$$

and

$$\frac{\partial^2 \xi_i}{\partial t^2} = -\frac{c_0^2}{\rho_0} \frac{\partial \rho}{\partial a_i} \tag{7.11}$$

since $p' = c_0^2 \rho'$. That is,

$$\frac{\partial^2 \rho}{\partial t^2} = c_0^2 \frac{\partial^2 \rho}{\partial a_i \partial a_i} \,, \tag{7.12}$$

which is a wave equation in the Lagrangian coordinates. At this order of accuracy, however, the Lagrangian and Eulerian rates of change for time are the same, as are those for space.

7.2 Hydrodynamic stability

Consider incompressible flow in the plane with Eulerian coordinates (x, y, t), Lagrangian coordinates $(a, b, s|t)$, and particle paths $(X, Y) = (a + U[b](t-s) + \xi, b + \eta)$.

Note that $U = U[y]$ is a steady Eulerian shear flow. The linearized Lagrangian equations of motion are:

$$\frac{\partial^2 \xi}{\partial t^2} = -\frac{1}{\rho}\frac{\partial p}{\partial a}, \tag{7.13}$$

$$\frac{\partial^2 \eta}{\partial t^2} = -\frac{1}{\rho}\mathcal{D}p \tag{7.14}$$

and

$$\frac{\partial \xi}{\partial a} + \mathcal{D}\eta = 0, \tag{7.15}$$

where

$$\mathcal{D} = \frac{\partial}{\partial b} - (t-s)\frac{dU}{db}\frac{\partial}{\partial a}. \tag{7.16}$$

It readily follows, without further approximation, that

$$\frac{\partial^4 \eta}{\partial a^2 \partial t^2} + \mathcal{D}\left(\frac{\partial^2 (\mathcal{D}\eta)}{\partial t^2}\right) = 0. \tag{7.17}$$

Let

$$\eta(a, b, s|t) = B\big(a + U[b](t-s), b, s|t\big) \tag{7.18}$$

(this is cheating with the mean flow) and assume that

$$B(a, b, s|t) = \frac{C(b)}{(\sigma + U[b]\kappa)}\exp\Big(i\big(\kappa a + \sigma(t-s)\big)\Big), \tag{7.19}$$

where $i = \sqrt{-1}$, and C and σ are complex valued. The integrating factor converts the transverse displacement amplitude B to the transverse velocity amplitude C – more cheating. Then

$$\frac{d^2 C}{db^2} - \left(\kappa^2 + \frac{\kappa}{\sigma + U\kappa}\frac{d^2 U}{db^2}\right)C = 0. \tag{7.20}$$

Assuming that

$$\alpha_0 \frac{dC}{db} + \beta_0 C \text{ vanishes at } b = b_0,$$

and

$$\alpha_1 \frac{dC}{db} + \beta_1 C \text{ vanishes at } b = b_1,$$

where $\alpha_{0,1}$ and $\beta_{0,1}$ are real, it follows that

$$\sigma_i \kappa \int_{b_0}^{b_1} \frac{|C|^2}{|\sigma + U\kappa|^2} \frac{d^2U}{db^2} db = 0, \tag{7.21}$$

where σ_i is the imaginary part of σ. Thus a necessary condition for instability ($\sigma_i < 0$) is that $U[b]$ has a point of inflexion in the range $b_0 < b < b_1$. Indeed, (7.20) is Rayleigh's stability equation for parallel shear flows. All the general criteria for instability hold, such as Howard's semicircle theorem (see, e.g., Drazin and Reid, 1981, Section 22).

The stability of infinitesimal disturbances to diverging flows or time-dependent flows may be investigated by considering only disturbances with space- and time-scales much smaller than those of the basic flow. Leblanc (2004) finds the Gerstner wave of Section 6.2 to be stable if and only if the 'steepness parameter' is small: $\exp(mb) < 1/3$.

The local kinematics and local dynamics, analysed in Section 1.3 and in Section 3.10, respectively, both indicate the possibility of particle displacements growing exponential in time, in *any* flow; thus the analysis of linearized hydrodynamic stability would seem uninformative or even contradictory. There is, however, a fundamental difference between the two analyses. In local kinematics and dynamics, the Lagrangian velocity field and Lagrangian enthalpy field, respectively, are expanded as Taylor series about the labeling position and labeling time, in powers of the particle displacement and the time elapsed since labeling. The coefficients in the Taylor series are derivatives of the Eulerian fields at the labeling position and time. In particular, the conclusions are only valid locally in space. There is no local expansion of the fields in either Eulerian or Lagrangian linearized hydrodynamic stability theory: the nonlinear dynamical equations are expanded in powers of field amplitude alone, and the conclusions are globally valid in space and time.

7.3 Rossby waves

Consider incompressible planar flow in a nonuniformly rotating reference frame. The particle labels are their initial Cartesian coordinates (a, b).

In Lagrangian form, the finite-amplitude equations for conservation of momentum and mass are:

$$\frac{\partial u}{\partial t} - fv = -\rho^{-1}\frac{\partial(p, Y)}{\partial(a, b)}, \tag{7.22}$$

$$\frac{\partial v}{\partial t} + fu = -\rho^{-1}\frac{\partial(X, p)}{\partial(a, b)}, \tag{7.23}$$

$$\frac{\partial(X, Y)}{\partial(a, b)} = 1, \tag{7.24}$$

where

$$u = \frac{\partial X}{\partial t}, \quad v = \frac{\partial Y}{\partial t}. \tag{7.25}$$

The Coriolis parameter is approximated as a linear function of the meridional (northward) coordinate:

$$f = f_0 + \beta(Y - y_0) \tag{7.26}$$

where f_0, y_0 and β are constants.

The Cauchy–Weber integrals are:

$$\frac{\partial X}{\partial a}u + \frac{\partial Y}{\partial a}v + \beta XY\frac{\partial Y}{\partial a} = -\frac{\partial \varphi}{\partial a} + u_s, \tag{7.27}$$

$$\frac{\partial X}{\partial b}u + \frac{\partial Y}{\partial b}v + \beta XY\frac{\partial Y}{\partial b} = -\frac{\partial \varphi}{\partial b} + v_s + \beta ab, \tag{7.28}$$

where

$$\varphi = \int_s^t \left(\frac{p}{\rho} - \gamma - \beta XY\frac{\partial Y}{\partial t} - \frac{1}{2}(u^2 + v^2)\right) dr, \tag{7.29}$$

while u_s and v_s are independent of time t. The time-dependent field γ is a potential[1] for the Coriolis acceleration associated with a reference value of the Coriolis parameter:

$$\frac{\partial \gamma}{\partial a} = -f_1\left(\frac{\partial X}{\partial a}v - \frac{\partial Y}{\partial a}u\right), \tag{7.30}$$

$$\frac{\partial \gamma}{\partial b} = -f_1\left(\frac{\partial X}{\partial b}v - \frac{\partial Y}{\partial b}u\right), \tag{7.31}$$

[1] For an alternative form of (7.27)–(7.29), see Section 16.2.2

where $f_1 = f_0 - \beta y_0$. The existence of the potential γ follows from the incompressibility condition (7.24). The Cauchy–Weber integration constants in (7.27) and (7.28) are:

$$u_s = u(a, b, s|s), \tag{7.32}$$

$$v_s + \beta ab = v(a, b, s|s) + \beta ab. \tag{7.33}$$

Assume that the particle motion is rectilinear, in the direction perpendicular to (κ, λ). That is, assume $\kappa u + \lambda v = 0$. It follows immediately that

$$\kappa X + \lambda Y = \kappa a + \lambda b \tag{7.34}$$

for all $t \geq s$. Assume also that the motion is a plane wave. That is, all Lagrangian variables are functions only of the phase θ where

$$\theta = \kappa X + \lambda Y + \sigma t = \kappa a + \lambda b + \sigma t, \tag{7.35}$$

where σ is some frequency. Then (7.22)–(7.26) reduce to

$$\frac{d^2 V}{d\theta^2} + \frac{\beta \kappa}{\sigma(\kappa^2 + \lambda^2)} V = 0, \tag{7.36}$$

where $v = V(\theta)$. We may infer without loss of generality that

$$V = V_0 \sin\theta, \tag{7.37}$$

where V_0 is a constant, and accordingly

$$\sigma = \frac{\beta \kappa}{(\kappa^2 + \lambda^2)}, \tag{7.38}$$

which is the dispersion relation for these *finite-amplitude* Rossby waves. The particle paths are

$$X = a + \frac{\lambda V_0}{\kappa \sigma}(\cos\theta - \cos\xi), \tag{7.39}$$

$$Y = b - \frac{V_0}{\sigma}(\cos\theta - \cos\xi), \tag{7.40}$$

where $\xi = \kappa a + \lambda b + \sigma s$.

Exercise 7.1 Verify that the particle paths (7.39), (7.40) satisfy the labeling theorem (1.4). □

Exercise 7.2 Show that the Cauchy–Weber integral scalar is

$$
\begin{aligned}
\varphi = &-\frac{\lambda V_0}{\kappa^2}(\cos\theta - \cos\xi) - \frac{\beta V_0 b}{\sigma\lambda}(\sin\theta - \sin\xi) \\
&- \frac{\beta V_0^2}{4\kappa\sigma^2}\Big(2\sin(\theta+\xi) - \sin(\theta+\xi)\cos(\theta-\xi) - \sin 2\theta\Big) \\
&+ \beta\left(\frac{\lambda}{6\kappa}(b^3 - 3bY^2 + 2Y^3) + \frac{a}{2}(b^2 - Y^2)\right) .
\end{aligned}
\tag{7.41}
$$

See Figure 7.1. The labeling velocities u_s, v_s are independent of f_0, so (7.27) and (7.28) ensure that φ is independent of f_0. Finally, (7.29) ensures that $p/\rho - \gamma$ is also independent of f_0. A uniform rotation of an incompressible planar flow is equivalent to a change of the pressure. □

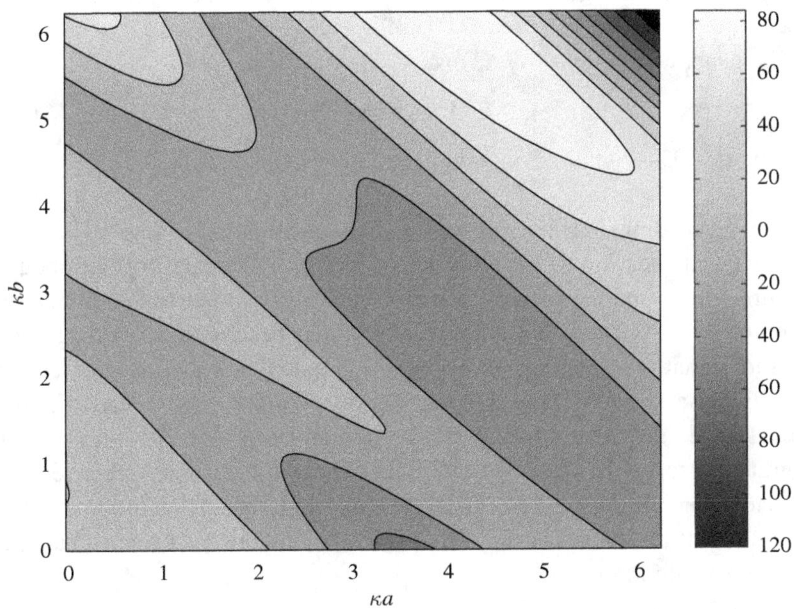

Figure 7.1 Plane Rossby wave: the scaled Cauchy–Weber integral scalar $\varphi' = V_0\varphi/\lambda$, as a function of the scaled labels κa and κb both in the range $[0, 2\pi]$. The dimensionless parameters are the Rossby number based on β: $R_\beta \equiv V_0\kappa\lambda/\beta$, and the aspect ratio: $r \equiv \kappa/\lambda$. Here, $R_\beta = 1$, $r = 1$, $\sigma t = 3\pi/4$, $s = 0$. A colour animation of this figure may be found at `http://www.cambridge.org/0521853109`

Exercise 7.3 Considering a general compressible planar flow with arbitrary labels, derive the following conservation law for the total Cauchy invariant:

$$\frac{\partial}{\partial t}(\varpi + fJ_s^t) = 0\,, \tag{7.42}$$

where ϖ is the Cauchy invariant relative to the rotating reference frame, fJ_s^t is the planetary Cauchy invariant, and J_s^t is the Jabobi determinant for the transformation $(a, b) \to (X, Y)$. Find ϖ, as a function of time t, in the incompressible plane Rossby wave (7.39), (7.40). □

Exercise 7.4 From (7.42) derive the Fofonoff (1954) equation for steady, incompressible inviscid ocean circulation on the β-plane:

$$\frac{\partial(\Psi, \boldsymbol{\nabla}^2\Psi + \beta b)}{\partial(a, b)} = 0\,, \tag{7.43}$$

where the Laplacian is taken with respect to a and b, and Ψ is a streamfunction for the initial Lagrangian velocities $u(a, b, s|s)$ and $v(a, b, s|s)$. Hint: Use the labeling theorem (1.4).

The general solution of (7.43) is

$$\boldsymbol{\nabla}^2\Psi + \beta b = F(\Psi) \tag{7.44}$$

(Fofonoff, 1954), where F is any differentiable function. □

The preceeding exercise illustrates a general rule for Lagrangian fluid dynamics. If the flow is steady, that is, if the flow fields depend upon the labeling time s and the time of interest t only in the time translation invariant combination $t - s$, then the labeling theorem may be used to derive Lagrangian equations that are identical in form to the Eulerian equations for conventionally defined steady flow. Taking into account the coalescence of particle paths, streaklines and streamlines, the unfamiliar complexity of alternative Lagrangian forms for steady flow makes those other forms seem less useful. However, while the Eulerian forms are comfortingly familiar, it can be very difficult to prove that they have smooth solutions, or even to attempt to compute a solution numerically.

Exercise 7.5 Assume that the variables $(a, b, t, X, Y, u, v, \varphi)$ in (7.27)–(7.29) have the respective scales $(\mathcal{A}, \mathcal{A}, \mathcal{T}, \mathcal{X}, \mathcal{X}, \mathcal{V}, \mathcal{V}, \mathcal{F})$. Show that the choice $\mathcal{F} = \beta\mathcal{X}^3$ reduces the equations to forms involving the dimensionless groups $\mathcal{A}/\mathcal{X}$ and $R_\beta = \mathcal{V}/(\beta\mathcal{X}^2)$. Infer that $R_\beta \gtrsim 1$ for a free nonzonal flow, and hence the scale $\mathcal{K} = \mathcal{V}\mathcal{X}$ for meridional diffusivities is bounded above: $\mathcal{K} \lesssim \sqrt{\mathcal{V}^3/\beta}$. □

7.4 Hamiltonian dynamics of Rossby waves

Following Chen and Byron-Scott (1995) and Dvorkin and Paldor (1999), consider the planar momentum equations (7.22), (7.23) as single-particle dynamics associated with the kinematics (7.25). The Coriolis parameter f is again given by (7.26). For clarity, suppose that $f_0 = 0$, that is, the particle is close to the equator. Assuming that the pressure field is steady and zonal, $p = p[Y]$, it is readily seen that there are two integrals of the motion:

$$\mathcal{U} = u - \frac{1}{2}\beta Y^2 , \tag{7.45}$$

and

$$\mathcal{E} = \frac{1}{2}(u^2 + v^2) + \frac{p}{\rho} . \tag{7.46}$$

Indeed, the system is Hamiltonian in the phase space having coordinates $(X, \mathcal{U}, Y, v)$, and the Hamiltonian function is

$$\mathcal{H}(X, \mathcal{U}, Y, v) = \frac{1}{2}\left(\mathcal{U} + \frac{1}{2}\beta Y^2\right)^2 + \frac{1}{2}v^2 + \frac{p}{\rho} . \tag{7.47}$$

Hamilton's equations are

$$\frac{\partial X}{\partial t} = \frac{\partial \mathcal{H}}{\partial \mathcal{U}} = \mathcal{U} + \frac{1}{2}\beta Y^2 , \tag{7.48}$$

$$\frac{\partial \mathcal{U}}{\partial t} = -\frac{\partial \mathcal{H}}{\partial X} = 0 , \tag{7.49}$$

$$\frac{\partial Y}{\partial t} = \frac{\partial \mathcal{H}}{\partial v} = v , \tag{7.50}$$

$$\frac{\partial v}{\partial t} = -\frac{\partial \mathcal{H}}{\partial Y} = -\left(\mathcal{U} + \frac{1}{2}\beta Y^2\right)\beta Y - \frac{p'[Y]}{\rho} . \tag{7.51}$$

The prime on p in (7.51) denotes a derivative.

If, for example, there is no cross-equatorial pressure gradient ($p' = 0$), then it is clear from (7.49) and (7.51) that the point $(X, \mathcal{U}, 0, 0)$ is the simply translating equilibrium state $X = a + \mathcal{U}(t - s)$, with eigenvalues $\pm\sqrt{-\beta\mathcal{U}}$. The equilibrium is stable if $\mathcal{U} > 0$, and unstable if $\mathcal{U} < 0$. If $\mathcal{U} < 0$ then there are also equilibria at $Y = \pm\sqrt{-2\mathcal{U}/\beta}$. Both are stable, thus there is a pitchfork bifurcation at $\mathcal{U} = 0$. Dvorkin and Paldor (1999) consider in detail the case of a nonvanishing cross-equatorial pressure gradient ($p' \neq 0$), and infer the general behavior of particle paths when the pressure is unsteady and not zonally uniform: $p = p[X, Y, t]$. The motion is no longer integrable and the paths are chaotic.

The preceding Hamiltonian reformulation and analysis of the Rossby wave momentum equations is certainly ingenious and fascinating, but of course the pressure gradient is not an external force. Pressure is an internal field which must be consistent with conservation of mass (7.24). Alternatively, the planar dynamics (7.22), (7.23) and kinematics (7.25) may be interpreted as shallow water theory, together with the Lagrangian continuity equation

$$\frac{1}{J_s^t}\frac{\partial(hJ_s^t)}{\partial t}=S \tag{7.52}$$

where the free surface height h is related to the barotropic pressure by $p=\rho gh$, g being the magnitude of the local gravitational acceleration, and where S is a fluid source. It follows that h can be steady and zonally uniform in the Eulerian sense, for variety of fluid sources rationalized as "heat sources".

7.5 Plane Ptolemaic vortices

Consider the Lagrangian formulation of unforced, inviscid incompressible flow in the plane:

$$\frac{\partial}{\partial t}\left(\frac{\partial\tilde{X}_1}{\partial a_1}\frac{\partial\tilde{X}_2}{\partial a_2}-\frac{\partial\tilde{X}_1}{\partial a_2}\frac{\partial\tilde{X}_2}{\partial a_1}\right)=0, \tag{7.53}$$

$$\frac{\partial}{\partial t}\left(\frac{\partial^2\tilde{X}_1}{\partial t\partial a_1}\frac{\partial\tilde{X}_1}{\partial a_2}+\frac{\partial^2\tilde{X}_2}{\partial t\partial a_1}\frac{\partial\tilde{X}_2}{\partial a_2}-\frac{\partial^2\tilde{X}_1}{\partial t\partial a_2}\frac{\partial\tilde{X}_1}{\partial a_1}-\frac{\partial^2\tilde{X}_2}{\partial t\partial a_2}\frac{\partial\tilde{X}_2}{\partial a_1}\right)=0, \tag{7.54}$$

where $(\tilde{X}_1(a_1,a_2,s|t),\tilde{X}_2(a_1,a_2,s|t))$ is the particle path.[2] However, it is *not* assumed that (a_1,a_2) is the position of the particle at time $t=s$.

Exercise 7.6 Introducing the complex variable $\eta=a_1+ia_2$ and the complex function $W=\tilde{X}_1+i\tilde{X}_2$, show that (7.53), (7.54) may be expressed as

$$\frac{\partial}{\partial t}\left(\left|\frac{\partial W}{\partial\eta}\right|^2-\left|\frac{\partial W}{\partial\overline{\eta}}\right|^2\right)=0, \tag{7.55}$$

$$\frac{\partial}{\partial t}\left(\frac{\partial^2 W}{\partial t\partial\eta}\frac{\partial\overline{W}}{\partial\overline{\eta}}-\frac{\partial^2 W}{\partial t\partial\overline{\eta}}\frac{\overline{W}}{\partial\eta}\right)=0, \tag{7.56}$$

where the overbar denotes the complex conjugate. Verify that

$$W=G(\eta)e^{i\lambda t}+F(\overline{\eta})e^{i\mu t}, \tag{7.57}$$

[2] The role of the tildes will become apparent in Section 7.6.

where F and G are arbitrary analytic functions, while λ and μ are arbitrary real numbers, is a solution of (7.55), (7.56). Find the Jacobi determinant and the Cauchy invariant. Show that the vorticity of the flow is the conserved scalar

$$\omega = \frac{2\left(\lambda\left|\frac{\partial G}{\partial \eta}\right|^2 - \mu\left|\frac{\partial F}{\partial \overline{\eta}}\right|^2\right)}{\left|\frac{\partial G}{\partial \eta}\right|^2 - \left|\frac{\partial F}{\partial \overline{\eta}}\right|^2}. \tag{7.58}$$

Are any of these solutions pathological? Show that the velocity field V satisfies the implicit relations

$$F^{-1}\left(e^{-i\mu t}\frac{V - i\lambda W}{i(\mu-\lambda)}\right) = \overline{G}^{-1}\left(e^{i\lambda t}\frac{\overline{V} + i\mu\overline{W}}{i(\mu-\lambda)}\right), \tag{7.59}$$

(Abrashkin and Yakubovich, 1984). Those authors remark: "It is obvious that for all the simplicity of the time dependence of the particle trajectories, the flow field in the Eulerian variables can be a very complicated function of time." □

The particle paths (7.57) are clearly circles, with centres that move in circles. That is, the paths are epicycloids. Yakubovich and Zenkovich (2001) are unable to resist the temptation to name these solutions "Ptolemaic vortices". The Cauchy–Weber integral scalar for the simple vortex $G(\eta) = A\eta$, $F(\overline{\eta}) = B\overline{\eta}$, where A and B are constants, is plotted in Figure 7.2.

The Gerstner trochoidal wave of Section 6.2 is a Ptolemaic vortex, with

$$G(\eta) = \eta, \quad \lambda = 0, \quad \mu = mc, \quad F(\overline{\eta}) = -\frac{i}{m}\exp(im\overline{\eta}). \tag{7.60}$$

A similar solution is

$$W = \zeta e^{i\lambda t} + F(\overline{\zeta}), \tag{7.61}$$

where $F(z)$ is any analytic function of a complex variable z. The conjugate of the complex fluid velocity is

$$\overline{V} = u - iv = -i\lambda\overline{\zeta}e^{-i\lambda t}. \tag{7.62}$$

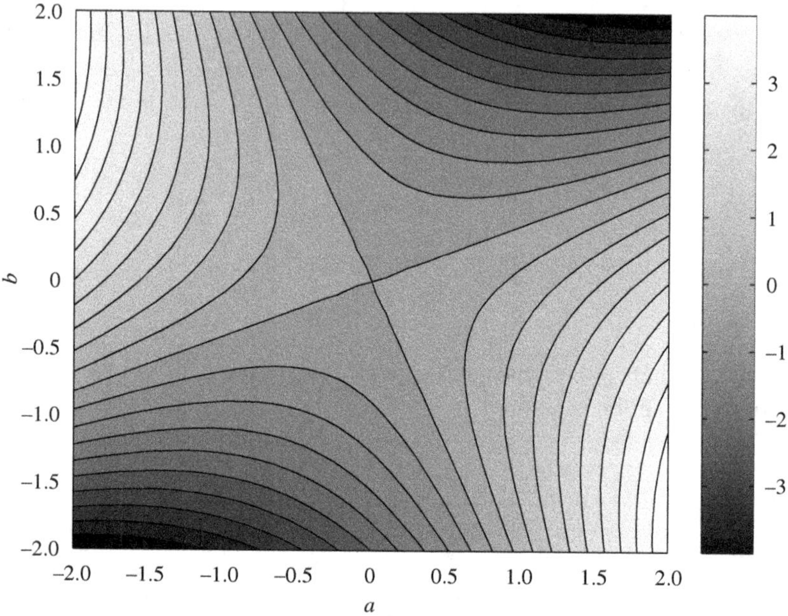

Figure 7.2 Plane Ptolemaic vortex: the scaled Cauchy–Weber integral scalar $\varphi/(\sigma AB)$ as a function of the real labels (a, b) at the scaled time $\sigma t = \pi/2$; the complex particle path is $W = X + iY = A\eta \exp(i\lambda t) + B\overline{\eta} \exp(i\mu t)$ with complex label $\eta = a + ib$, while σ is the frequency difference $\lambda - \mu$. A colour animation of this figure may be found at `http://www.cambridge.org/0521853109`

Note that at $t = 0$, $W = \zeta + F(\overline{\zeta})$. Thus ζ is a complex parameter for the initial position of the fluid particles. Let this Ptolemaic vortex be defined for ζ within the unit circle: $|\zeta| \leq 1$. Then for ζ on this circle, say for $\zeta = e^{i\phi}$:

$$W = e^{i\phi} e^{i\lambda t} + F(e^{-i\phi}) , \tag{7.63}$$

$$\overline{V} = u - iv = -i\lambda e^{-i\phi} e^{-i\lambda t} . \tag{7.64}$$

Introduce a new complex variable $\xi = Ae^{i\phi}$, where $A \geq 0$. Following Abrashkin and Yakubovich (1984), for $A \geq 1$ let the complex particle path be

$$W = \xi e^{i\lambda t} + F(\xi^{-1}) , \tag{7.65}$$

and let the conjugate complex velocity be

$$\overline{V} = u - iv = -i\lambda \xi^{-1} e^{-i\lambda t} . \tag{7.66}$$

For each value of t, (7.65) may in principle be inverted to yield ξ as an analytic function of W. Thus $\overline{V}$ is an analytic function of W. The Cauchy–Riemann conditions therefore apply to the real and imaginary parts of $\overline{V}$; accordingly the Eulerian fluid velocity $(u[x, y, t], v[x, y, t])$ is both solenoidal and irrotational. So (7.65), (7.66) constitute a parametric representation of a potential flow that has the same velocity (both components) as the Ptolemaic vortex (7.61) at the boundary (7.63).

Exercise 7.7 Verify that the pressure is continuous across the boundary between the Ptolemaic vortex and the potential flow. Find a condition on F ensuring nonsingularity of the vortex. Estimate the asymptotic decay rate of the potential flow decay far from the vortex. □

Exercise 7.8 Abrashkin and Yakubovich (1984): choose F in (7.61) to be $F(z) = -\alpha z + \beta z^2$, where $|\alpha| + 2|\beta| < 1$. Plot the boundary of the vortex and the streamlines on either side, for $\lambda(t-s) = 0, \pi/4, \ldots, 7\pi/4$. □

Exercise 7.9 How might the Ptolemaic vortex be extended to labeling times other than $s = 0$? □

7.6 Sheared Ptolemaic vortices

The class of planar vortices in incompressible flow examined in the preceding section is extended to three-dimensional motions by Yakubovich and Zenkovich (2001, 2002). They are aided by the Lagrangian formulation involving the Jacobi matrix (see Section 3.7.2), which they term "matrix fluid dynamics." To begin, their extension will be given for an arbitrary solution of the planar Lagrangian equations of motion.

Let $\left(\tilde{X}_1(a_1, a_2, s|t), \tilde{X}_1(a_1, a_2, s|t)\right)$ be the particle paths for an exact solution of the inviscid equations of motion in the plane. The elements of the two-dimensional Jacobi matrix are as always

$$\tilde{J}_{ij} = \frac{\partial}{\partial a_j} \tilde{X}_i(a_1, a_2, s|t) \tag{7.67}$$

for $i = 1, 2$ and $j = 1, 2$. They may be extended to a Jacobi matrix J_{ij} in three dimensions as follows:

$$J_{ij} = \sqrt{f(t)}^{-1} \tilde{J}_{ij} \left(a_1, a_2, s| \int_s^t f(r) dr \right), \tag{7.68}$$

for $1=1,2$ and $j=1,2$;

$$J_{i3}=0\,, \tag{7.69}$$

for $i=1,2$;

$$J_{3j}=f(t)\left(\int_s^t f(r)^{-2}dr\right)\frac{\partial}{\partial a_j}H(a_1,a_2)\,, \tag{7.70}$$

where $j=1,2$;

$$J_{33}=f(t)\,, \tag{7.71}$$

where H is an arbitrary function of the planar coordinates (a_1,a_2) and $f(t)$ is an arbitrary function of time. The extension satisfies the consistency conditions (3.66) for a Jacobi matrix; indeed, the corresponding particle paths are

$$X_i(a_1,a_2,a_3,s|t)=\sqrt{f(t)}^{-1}\tilde{X}_i\left(a_1,a_2,s|\int_s^t f(r)dr\right)\,, \tag{7.72}$$

for $i=1,2$ and

$$X_3(a_1,a_2,a_3,s|t)=f(t)\left(H(a_1,a_2)\int_s^t f(r)^{-2}dr+a_3\right)\,. \tag{7.73}$$

The flow is three-dimensional, with vertical velocity

$$u_3=\left(H+\frac{df}{dt}X_3\right)f^{-1}\,. \tag{7.74}$$

The extended Jacobi determinant is

$$\Delta\equiv\det\Big(J_{ij}(a_1,a_2,a_3,s|t)\Big)=\det\left(\tilde{J}_{ij}(a_1,a_2,s|\int_s^t f(r)dr)\right)\,. \tag{7.75}$$

If the right-hand side is independent of time t, then so is the left-hand side. That is, if the planar flow is incompressible, then so is this extension to three dimensions. In any event, the Cauchy invariants are

$$\varpi_1=\frac{\partial H}{\partial a_2},\qquad \varpi_2=-\frac{\partial H}{\partial a_1},\qquad \varpi_3=\tilde{\varpi}\,, \tag{7.76}$$

where $\tilde{\varpi}$ is the planar Cauchy invariant:

$$\tilde{\varpi}=\frac{\partial\tilde{J}_{11}}{\partial t}\tilde{J}_{12}+\frac{\partial\tilde{J}_{21}}{\partial t}\tilde{J}_{22}-\frac{\partial\tilde{J}_{12}}{\partial t}\tilde{J}_{11}-\frac{\partial\tilde{J}_{22}}{\partial t}\tilde{J}_{21}\,. \tag{7.77}$$

Note that assuming $(\tilde{X}, \tilde{Y})$ is a planar incompressible solution is equivalent to assuming that $\det(\tilde{J}_{ij})$ and $\tilde{\varpi}$ are invariants. Thus the vector ϖ_i is indeed invariant; however, the vorticity ω_i is time-dependent:

$$\omega_1 = f^{-1}\frac{\partial H}{\partial X_2}, \qquad \omega_2 = -f^{-1}\frac{\partial H}{\partial X_1}, \qquad \omega_3 = f\tilde{\varpi}\Delta^{-1}. \tag{7.78}$$

Exercise 7.10 Derive (7.78) using (3.69); display the space and time arguments in detail. Colour animations of examples of sheared Ptolemaic vortices may be found at `http://www.cambridge.org/0521853109` □

8
Viscous incompressible flow

8.1 Simple shear flow

Consider viscous incompressible flow between parallel plates occupying the planes $y=0$ and $y=H$. The first plate is fixed, while the second slides in the Ox direction with speed U_H. There is a uniform Eulerian pressure gradient Γ in that same direction. It may therefore be assumed that the fluid velocity is $(u, 0, 0)$, and

$$\frac{\partial u}{\partial a}=\frac{\partial u}{\partial c}=0. \tag{8.1}$$

Thus $u=u(b, s|t)$, where (a, b, c) is the initial position of a fluid particle. It follows that the particle path is

$$(X, Y, Z)=\left(a+\int_s^t u(b, s|r)dr, b, c\right). \tag{8.2}$$

Conservation of mass is trivially satisfied:

$$J=\frac{\partial(X, Y, Z)}{\partial(a, b, c)}=1. \tag{8.3}$$

The only nontrivial component of conservation of momentum is (5.4), which becomes

$$\frac{\partial u}{\partial t}=-\frac{1}{\rho}\Gamma+\nu\frac{\partial^2 u}{\partial b^2}. \tag{8.4}$$

Note that

$$\frac{\partial p}{\partial x}=\frac{\partial p}{\partial a}=\Gamma.$$

Assuming that the flow is steady, that is $u(b, s|t) = u(b|t-s)$, it follows that

$$\frac{\partial u}{\partial t} = -\frac{\partial u}{\partial s}.$$

But $\partial u/\partial s = 0$ as a consequence of the labeling theorem (1.4) and the downstream uniformity of the flow. So, after some simple calculus and imposing no-slip boundary conditions at $Y = b = 0$ and at $Y = b = H$:

$$u = \frac{\Gamma}{2\rho\nu}(b^2 - bH) + U_H \frac{b}{H}, \tag{8.5}$$

while $X = a + (t-s)u$, $Y \equiv b$ and $Z \equiv c$. If the second plate is also at rest: $U_H = 0$, then the fluid velocity has its extreme value $u_{\max} = -\Gamma H^2/(8\rho\nu)$ midway between the plates ($b = 0.5H$). Construction of the solution in Lagrangian coordinates is virtually identical to the Eulerian construction, but the former invokes the conservation of particle identity.

8.2 The suddenly accelerated plane wall: Stokes' first problem

A flat plate occupies the plane $y = 0$, and is suddenly accelerated from rest to slide with uniform velocity $(U_0, 0, 0)$ through an incompressible viscous fluid. The only nontrivial equation of motion is again

$$\frac{\partial u}{\partial t} = \nu \frac{\partial^2 u}{\partial b^2}, \tag{8.6}$$

subject to the boundary conditions

$$t \leq 0: u = 0 \text{ for } b = 0; \quad t > 0: u = U_0 \text{ for } b = 0; \quad u = 0 \text{ as } |b| \to \infty. \tag{8.7}$$

The solution of this textbook diffusion problem is readily shown to be

$$u(b, 0|t) = U_0 \operatorname{erfc}\left\{\frac{b}{2\sqrt{\nu t}}\right\} \tag{8.8}$$

where erfc is the complementary error function. Note that the solution is a boundary layer of width $2\sqrt{\nu t}$. As $t \to \infty$, the entire fluid eventually attains the velocity of the plate: see Figure 8.1.

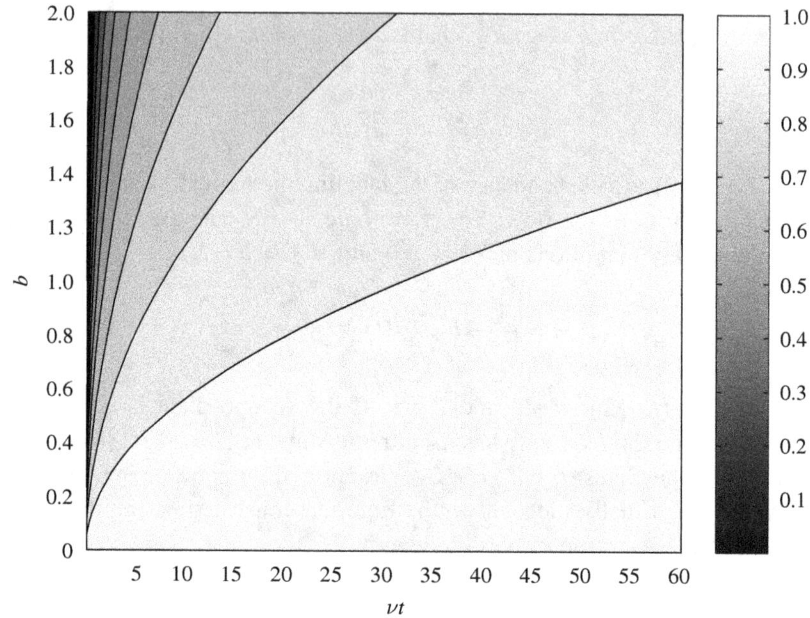

Figure 8.1 Stoke's first problem: the scaled velocity u/U_0 as a function of scaled time νt and normal distance b.

8.3 Flow near an oscillating flat plate: Stokes' second problem

Now suppose that the sliding of the plate is oscillatory: $u = U_0 \cos(\sigma t)$ at $b = 0$, for some real frequency σ. The solution is easily seen to be

$$u(b, 0|t) = U_0 \exp(-kb) \cos(\sigma t - kb) \tag{8.9}$$

where $k = \sqrt{\sigma/2\nu}$. The boundary layer width k^{-1} is a constant, and the oscillations of the fluid lag those of the plate by kb radians. Unlike the preceding solution for Stokes' first problem, the particle paths are explicitly available:

$$X(b, s|t) = a + \frac{U_0}{\sigma} \exp(-kb) \Big(\sin(\sigma t - kb) - \sin(\sigma s - kb) \Big). \tag{8.10}$$

These rectilinear orbits are closed, that is, there is no Stokes' drift: see Figure 8.2.

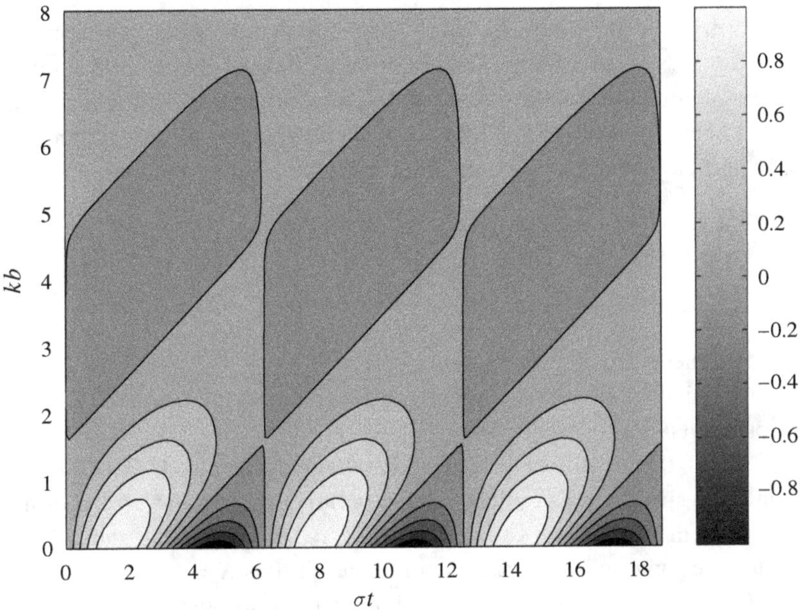

Figure 8.2 Stoke's second problem: the scaled displacement $(X-a)/(U_0/\sigma)$ as a function of scaled time σt and scaled normal distance kb where the boundary layer width is $k^{-1}=\sqrt{2\nu/\sigma}$.

8.4 The boundary layer along a flat plate

A fixed plate occupies the half-plane $y=0,\ x>0$; the fluid velocity in the far field has the asymptote $(U_\infty, 0, 0)$ as $y\to\infty$, where U_∞ is a constant. There is no pressure gradient in the far field. Following Prandtl's scale analysis near the plate (Schlichting, 1960), and using the labeling theorem in the case of steady flow, the boundary layer approximation to the Lagrangian form of the equations of motion are

$$\frac{\partial^2 X}{\partial t^2}=\nu\left(\frac{\partial(X,\ )}{\partial(a,b)}\right)^2\frac{\partial X}{\partial t} \tag{8.11}$$

where $X=X(a,b|t-s)$, $Y=Y(a,b|t-s)$ and

$$\frac{\partial X}{\partial a}\frac{\partial Y}{\partial b}-\frac{\partial X}{\partial b}\frac{\partial Y}{\partial a}=1\,. \tag{8.12}$$

These Lagrangian equations are substantially more complex than the Eulerian forms. However, setting $t=s$ and using the labeling theorem reduces the Lagrangian equations to the Eulerian forms for the Eulerian velocities. These latter forms may then be attacked by the introduction of a similarity variable $\eta = b\sqrt{U_\infty/(\nu a)}$, and by numerical integration of the resulting ordinary differential equation for the dimensionless streamfunction, following Blasius (see Schlichting, 1960). The Eulerian velocities having been computed, the steady particle paths may be computed from the labeling theorem:

$$\frac{\partial X}{\partial t} = u[a,b]\frac{\partial X}{\partial a} + v[a,b]\frac{\partial X}{\partial b}, \tag{8.13}$$

$$\frac{\partial Y}{\partial t} = u[a,b]\frac{\partial Y}{\partial a} + v[a,b]\frac{\partial Y}{\partial b}. \tag{8.14}$$

Note that these are *linear* equations for $X(a,b|t-s)$, $Y(a,b|t-s)$, albeit with coefficients that vary in labeling space. They express no more than the relationship (1.17) for steady flow.

But all of that is cheating. The direct Lagrangian attack is to introduce not only the similarity variable $\eta = b\sqrt{U_\infty/(\nu a)}$ for the transverse coordinate b, but also the similarity variable $\tau = tU_\infty/a$ for the time t and a scaling by a for the longitudinal particle coordinate X: thus $X = a\xi(\eta|\tau-\sigma)$ where $\sigma = sU_\infty/a$. The longitudinal momentum equation becomes

$$\frac{\partial^2\xi}{\partial\tau^2} = P\frac{\partial Q}{\partial\eta} + \frac{\partial\xi}{\partial\eta}\left\{\frac{1}{2}\frac{\partial}{\partial\eta}(\eta Q) + (\tau-\sigma)\frac{\partial Q}{\partial\tau}\right\} \tag{8.15}$$

where

$$P = \xi - \frac{1}{2}\eta\frac{\partial\xi}{\partial\eta} - (\tau-\sigma)\frac{\partial\xi}{\partial\tau}, \quad Q = \xi\frac{\partial^2\xi}{\partial\eta\partial\tau} - (\tau-\sigma)\left(\frac{\partial\xi}{\partial\tau}\frac{\partial^2\xi}{\partial\eta\partial\tau} - \frac{\partial\xi}{\partial\eta}\frac{\partial^2\xi}{\partial\tau^2}\right). \tag{8.16}$$

Thus the momentum equation has reduced to a partial differential equation for $\xi = \xi(\eta|\tau-\sigma)$, subject to $\xi = 1$ at $\tau = \sigma$ and $\partial\xi/\partial\tau \sim 1$ as $\eta \to \infty$. The solution for ξ and hence X having been found, the solution for $Y = b + \sqrt{\nu a/U_\infty}\,\phi(\eta|\tau-\sigma)$ is found from the continuity equation:

$$\left(1 + \frac{\partial\phi}{\partial\eta}\right)P - \frac{\partial\xi}{\partial\eta}\left\{\frac{1}{2}\phi - \frac{1}{2}\eta\frac{\partial\phi}{\partial\eta} - (\tau-\sigma)\frac{\partial\phi}{\partial\tau}\right\} = 1 \tag{8.17}$$

subject to $\phi = 0$ at $\tau = \sigma$ and $\phi \sim 0$ as $\eta \to \infty$.

Exercise 8.1 Explain the use of the labeling theorem for scale analysis of steady flow. □

Exercise 8.2 Solve (8.15)–(8.17) numerically, plotting profiles of the similarity functions much as in Schlichting (1960). □

9

General solvability

The general solvability for the Lagrangian dynamics of an inviscid, non-conducting fluid is examined in some detail. Some remarks are made about viscous conducting fluids.

9.1 Kinematics

It is customary to compute particle paths from Eulerian numerical solutions by integrating the Eulerian velocity field along a path. The defining relationship

$$\frac{\partial X_i}{\partial t} = u_i[X_j, t], \tag{9.1}$$

subject to $X_i = a_i$ at $t = s$, is integrated numerically to obtain the paths $X_i(a_j, s|t)$. This relationship is the subject of much local analysis, see for example Batchelor (1952), Ottino (1989) or Samelson and Wiggins (2005). Assuming that the label a_i is the particle position at time s, the nonlinear relationship (9.1) is expanded in space as a Taylor series:

$$\frac{\partial}{\partial t} X_i(a_j, s|t) = u_i[a_j, t] + \frac{\partial u_i}{\partial x_k}[a_j, t]\big(X_k(a_j, s|t) - a_k\big) + \cdots \tag{9.2}$$

The system of linear ordinary differential equations (9.2) has a formal solution in terms of exponentiated time integrated matrices; the essential point here is that this kinematical problem is well posed.

9.2 Incompressible dynamics (1)

The dynamics of incompressible flow may be expressed as partial differential equations for the particle paths. For simplicity assume a planar flow of particles, labeled by their release positions (a, b), on paths $(X, Y) = (X(a, b, s|t), Y(a, b, s|t))$; the particle velocities are $(u, v) = (u(a, b, s|t), v(a, b, s|t))$. The Cauchy–Weber integral relation (3.46), and scalar (3.47) become

$$\frac{\partial X}{\partial a}u + \frac{\partial Y}{\partial a}v = -\frac{\partial \varphi}{\partial a} + U \equiv P, \tag{9.3}$$

$$\frac{\partial X}{\partial b}u + \frac{\partial Y}{\partial b}v = -\frac{\partial \varphi}{\partial b} + V \equiv Q, \tag{9.4}$$

where

$$\varphi = \int_s^t \left\{ h + \mathcal{F} - \frac{1}{2}(u^2 + v^2) \right\} dr \tag{9.5}$$

and

$$U = u(a, b, s|s), \quad V = v(a, b, s|s). \tag{9.6}$$

Recall that h is the enthalpy, and $\mathcal{F}$ is the potential for an irrotational body force. For planar incompressible flow, conservation of mass is equivalent to conservation of area:

$$\frac{\partial}{\partial t}\left(\frac{\partial X}{\partial a}\frac{\partial Y}{\partial b} - \frac{\partial X}{\partial b}\frac{\partial Y}{\partial a} \right) = 0. \tag{9.7}$$

The linear system (9.3), (9.4) is readily solved for (u, v):

$$\begin{bmatrix} u \\ v \end{bmatrix} = \begin{bmatrix} \dfrac{\partial Y}{\partial b} & -\dfrac{\partial Y}{\partial a} \\ -\dfrac{\partial X}{\partial b} & \dfrac{\partial X}{\partial a} \end{bmatrix} \begin{bmatrix} P \\ Q \end{bmatrix} \tag{9.8}$$

or

$$\frac{\partial}{\partial t}\begin{bmatrix} X \\ Y \end{bmatrix} = \begin{bmatrix} 0 & -Q \\ Q & 0 \end{bmatrix} \frac{\partial}{\partial a}\begin{bmatrix} X \\ Y \end{bmatrix} + \begin{bmatrix} 0 & P \\ -P & 0 \end{bmatrix} \frac{\partial}{\partial b}\begin{bmatrix} X \\ Y \end{bmatrix}. \tag{9.9}$$

This system of first-order equations may be decoupled, yielding the second-order equations

$$\frac{\partial^2}{\partial t^2}\begin{bmatrix} X \\ Y \end{bmatrix} = \left(-Q^2\frac{\partial^2}{\partial a^2} + 2PQ\frac{\partial^2}{\partial a \partial b} - P^2\frac{\partial^2}{\partial b^2} \right)\begin{bmatrix} X \\ Y \end{bmatrix} + \cdots \tag{9.10}$$

where the ellipsis denotes derivatives of (X, Y) of lower order. The operator in (9.10) cannot be characterized until the dynamics of P and Q have been developed. These fields are defined in (9.3) and (9.4); they depend upon the initial velocity (U, V) and the Cauchy–Weber integral scalar $\varphi(a, b, s|t)$. Substituting (9.9) into the incompressibility condition (9.7) yields

$$\left[\left(\frac{\partial X}{\partial b}\right)^2+\left(\frac{\partial Y}{\partial b}\right)^2\right]\frac{\partial^2\varphi}{\partial a^2}-2\left[\frac{\partial X}{\partial a}\frac{\partial X}{\partial b}+\frac{\partial Y}{\partial a}\frac{\partial Y}{\partial b}\right]\frac{\partial^2\varphi}{\partial a\partial b}$$
$$+\left[\left(\frac{\partial X}{\partial a}\right)^2+\left(\frac{\partial Y}{\partial a}\right)^2\right]\frac{\partial^2\varphi}{\partial b^2}+\cdots$$
$$=\left[\left(\frac{\partial X}{\partial b}\right)^2+\left(\frac{\partial Y}{\partial b}\right)^2\right]\frac{\partial U}{\partial a}-\left[\frac{\partial X}{\partial a}\frac{\partial X}{\partial b}+\frac{\partial Y}{\partial a}\frac{\partial Y}{\partial b}\right]\left(\frac{\partial V}{\partial a}+\frac{\partial U}{\partial b}\right)$$
$$+\left[\left(\frac{\partial X}{\partial a}\right)^2+\left(\frac{\partial Y}{\partial a}\right)^2\right]\frac{\partial V}{\partial b}+\cdots \tag{9.11}$$

where the ellipsis on the left-hand side denotes lower order derivatives of φ having coefficients that are second-order derivatives of X and Y, while the ellipsis on the right-hand side denotes analogous terms in U and V. It is readily verified that the operator on the left-hand side of (9.11) acting on φ is elliptic in (a, b). So, given suitable boundary conditions, (9.11) may be solved for φ whenever the particle paths (X, Y) are known. In particular, when $t=s$ the right-hand side vanishes and hence φ vanishes, subject to suitable boundary conditions. This is consistent with the definition (9.5) for φ. A trivial solution of (9.11) is

$$\left(\frac{\partial\varphi}{\partial a}, \frac{\partial\varphi}{\partial b}\right)=(U, V), \tag{9.12}$$

that is, $(P, Q)=(0, 0)$ in which case $(u, v)=(0, 0)$. Nontrivial solutions are supported by nonzero boundary values of (P, Q). The inference from (9.11), relevant to the characterization of (9.10), is that the value of φ at any point in (a, b, t) space does not depend on *local* values of second derivatives of (X, Y). Rather, it depends upon them globally, through their effect on the influence function for the elliptic φ-operator on the left-hand side of (9.11). Thus P and Q appearing in (9.10) may be regarded as the coefficients of a linear operator in (a, b, t) space. It is then immediately apparent that the operator is elliptic in (a, b, t) space. Thus the initial value problem for (9.10) with initial values $(X, Y)=(a, b)$, and

$$\left(\frac{\partial X}{\partial t}, \frac{\partial Y}{\partial t}\right)=(U, V)$$

is classically ill-posed (Hadamard, 1923, Section 18, p. 34). To paraphrase Hadamard, assume initial velocity fields of the form $\exp\left[i(\kappa a+\lambda b)\right]$. Then the time dependence will be of the form $\exp(\zeta t)$, where

$$\zeta^2=(Q\kappa-P\lambda)^2, \tag{9.13}$$

assuming that P and Q are slowly varying. Certain initial conditions will excite the mode that grows like

$$\left[\begin{array}{c} X \\ Y \end{array}\right] \sim \exp\left(i(\kappa a+\lambda b)+|Q\kappa-P\lambda|\,(t-s)\right). \tag{9.14}$$

Note that the growth rate in (9.14) is not simply the local rate of strain, as in the kinematic analysis (9.2), but rather is proportional to the initial wavenumber $\sqrt{\kappa^2+\lambda^2}$. Thus, as pointed out by Hadamard, an initially smooth field would rapidly become "fluted."

9.3 Incompressible dynamics (2)

The Lagrangian formulation based on the Cauchy–Weber integral relations (9.3), (9.4) and the continuity equation (9.7) is ill-posed, but there are alternative formulations. Again for planar flow, the momentum equations in the absence of a body force are

$$\rho\frac{\partial^2 X}{\partial t^2}=-\frac{\partial p}{\partial a}\frac{\partial Y}{\partial b}+\frac{\partial p}{\partial b}\frac{\partial Y}{\partial a}, \tag{9.15}$$

$$\rho\frac{\partial^2 Y}{\partial t^2}=-\frac{\partial X}{\partial a}\frac{\partial p}{\partial b}+\frac{\partial X}{\partial b}\frac{\partial p}{\partial a}. \tag{9.16}$$

Substituting (9.15), (9.16) into the time derivative of the incompressibility condition (9.7) yields a pressure equation that is elliptic in (a, b) space:

$$\left[\left(\frac{\partial X}{\partial b}\right)^2+\left(\frac{\partial Y}{\partial b}\right)^2\right]\frac{\partial^2 p}{\partial a^2}-2\left[\frac{\partial X}{\partial a}\frac{\partial X}{\partial b}+\frac{\partial Y}{\partial a}\frac{\partial Y}{\partial b}\right]\frac{\partial^2 p}{\partial a\partial b} +\left[\left(\frac{\partial X}{\partial a}\right)^2+\left(\frac{\partial Y}{\partial a}\right)^2\right]\frac{\partial^2 p}{\partial b^2}=\cdots \tag{9.17}$$

where the ellipsis denotes terms of lower order. It may be inferred that values of p at points in (a, b) space, at any time t, do not depend upon local values of first derivatives of X and Y. Thus (9.15), (9.16) may be characterized as a system of linear equations. Decoupling leads to

$$\frac{\partial^4}{\partial t^4}\begin{bmatrix} X \\ Y \end{bmatrix} = -\left(\frac{\partial p}{\partial a}\frac{\partial}{\partial b} - \frac{\partial p}{\partial b}\frac{\partial}{\partial a}\right)^2 \begin{bmatrix} X \\ Y \end{bmatrix} + \dots. \tag{9.18}$$

This is not a classical equation, but its character may again be determined by following Hadamard (1923, Section 18, p. 34). Assume initial velocity fields of the form $\exp\left[i(\kappa a + \lambda b)\right]$. Then the time dependence will be of the form $\exp(\zeta t)$, where

$$\zeta^4 = \left(\frac{\partial p}{\partial a}\lambda - \frac{\partial p}{\partial b}\kappa\right)^2, \tag{9.19}$$

assuming that p is slowly varying. Certain choices for initial conditions will excite the mode that grows like

$$\begin{bmatrix} X \\ Y \end{bmatrix} \sim \exp\left(i(\kappa a + \lambda b) + \left|\frac{\partial p}{\partial a}\lambda - \frac{\partial p}{\partial b}\kappa\right|^{\frac{1}{2}} (t - s)\right). \tag{9.20}$$

Note again that the growth rate in (9.20) is not simply the local rate of strain, as in the kinematic analysis (9.2), but rather is proportional to the square root of the initial wavenumber $\sqrt{\kappa^2 + \lambda^2}$. Thus, as pointed out by Hadamard (1923), an initially smooth field would fairly rapidly become "fluted."

9.4 Incompressible dynamics (3)

In a last-ditch attempt to find a well-posed Lagrangian formulation for incompressible flow, consider the Cauchy invariant (3.51). For planar fluid dynamics, this becomes the scalar

$$\alpha\frac{\partial X}{\partial t} + \beta\frac{\partial Y}{\partial t} \equiv \varpi = \varpi_s, \tag{9.21}$$

where

$$\alpha = \frac{\partial X}{\partial b}\frac{\partial}{\partial a} - \frac{\partial X}{\partial a}\frac{\partial}{\partial b}, \qquad \beta = \frac{\partial Y}{\partial b}\frac{\partial}{\partial a} - \frac{\partial Y}{\partial a}\frac{\partial}{\partial b} \tag{9.22}$$

and $\varpi_s = \varpi(a, b, s|s)$ is the invariant at time $t = s$. Conservation of mass may be expressed as

$$\beta\frac{\partial X}{\partial t} - \alpha\frac{\partial Y}{\partial t} = 0. \tag{9.23}$$

Decoupling (9.21) and (9.23) leads immediately to

$$(\alpha^2+\beta^2)\begin{bmatrix}\dfrac{\partial X}{\partial t}\\ \dfrac{\partial Y}{\partial t}\end{bmatrix}=\ldots, \tag{9.24}$$

where the ellipsis denotes first-order spatial partial derivatives. These equations for the particle velocity are elliptic in space, and accordingly are well-posed problems.

The above argument may be extended to three dimensions. The Cauchy invariant vector (3.51) may be organized as

$$M_{ij}\frac{\partial X_j}{\partial t}=\varpi_i\,, \tag{9.25}$$

where M_{ij} is a spatial differential operator that, for the purpose of characterizing the problem, may be regarded as linear and having constant coefficients. Similarly the incompressibility condition may be organized as

$$L_k\frac{\partial X_k}{\partial t}=0\,, \tag{9.26}$$

where L_k is a spatial differential operator that may also be regarded as linear with constant coefficients. It is obvious that M_{ij} must have the left null vector $\partial/\partial a_i$ and that this vector must be orthogonal to ϖ_i, since a curl has no divergence. The matrix M_{ij} must therefore also have a right null vector ρ_i. The system (9.25), (9.26) is uniquely solvable only if ρ_i is not orthogonal to L_i. That is, the spatial partial differential equation

$$L_k\rho_k\psi=0 \tag{9.27}$$

must have only the trivial solution $\psi(a_j)=0$. It may be shown that in fact $\rho_i=L_i$, thus (9.27) is the elliptic equation

$$L_kL_k\psi=0\,. \tag{9.28}$$

Assuming that homogeneous boundary conditions preclude nontrivial solutions, it may be concluded that the conservation laws for the Cauchy invariant vector and the Jacobi determinant in incompressible flow constitute a formally well-posed problem.

In conclusion, the Lagrangian form of incompressible fluid dynamics is formally well-posed, so long as the dynamics are expressed as a conservation law for the Cauchy invariant. The existence of smooth, or "classical" solutions has yet to be established.

9.5 Compressible dynamics

Conservation of mass implies

$$\rho = \rho_s J^{-1}, \tag{9.29}$$

where $\rho = \rho(a_i, s|t)$, and for clarity $\rho_s = \rho(a_i, s|s)$, $J = J_s^t(a_i, s|t)$. There is no loss of generality here by assuming that the Lagrangian labels a_i are the release positions. In particular, $J = 1$ at $t = s$. Consider next the conservation of energy (3.15); for an ideal gas, $\mathfrak{E} = C_\rho T$ and $p = R\rho T$, where C_ρ is the specific heat at constant density, T is the absolute temperature and R is the gas constant. Accordingly, (3.15) is equivalent to

$$\frac{\partial p}{\partial t} = \frac{\gamma p}{\rho}\frac{\partial \rho}{\partial t} \tag{9.30}$$

where $\gamma = C_p/C_\rho = (R + C_\rho)/C_\rho$ is the ratio of specific heats. It follows that

$$\frac{\partial \eta}{\partial t} = 0, \tag{9.31}$$

where

$$\eta = \eta_s + C_\rho \ln\left\{\frac{p}{p_s}\left(\frac{\rho}{\rho_s}\right)^{-\gamma}\right\} \tag{9.32}$$

is the entropy. Combining (9.29) and (9.32) yields

$$p = p_s J^{-\gamma}. \tag{9.33}$$

Consider now the momentum equations, in two dimensions for simplicity:

$$\rho\frac{\partial^2 X}{\partial t^2} = -\frac{\partial p}{\partial x} = -\frac{\partial(p, y)}{\partial(x, y)} = -\frac{\partial(A, B)}{\partial(x, y)}\frac{\partial(p, Y)}{\partial(a, b)} = -J^{-1}\frac{\partial(p, Y)}{\partial(a, b)}, \tag{9.34}$$

$$\rho\frac{\partial^2 Y}{\partial t^2} = -\frac{\partial p}{\partial y} = -\frac{\partial(x, p)}{\partial(x, y)} = -\frac{\partial(A, B)}{\partial(x, y)}\frac{\partial(X, p)}{\partial(a, b)} = -J^{-1}\frac{\partial(X, p)}{\partial(a, b)}, \tag{9.35}$$

since

$$J = \frac{\partial(X, Y)}{\partial(a, b)}. \tag{9.36}$$

Substituting (9.33) into (9.34) and (9.35) yields

$$\frac{\partial^2 X}{\partial t^2} = c_s^2 J^{-\gamma-1}\frac{\partial(J, Y)}{\partial(a, b)}, \tag{9.37}$$

$$\frac{\partial^2 Y}{\partial t^2} = c_s^2 J^{-\gamma-1}\frac{\partial(X, J)}{\partial(a, b)}, \tag{9.38}$$

where $c_s^2 = \gamma p_s/\rho_s$ is the sound speed in an isentropic gas. Note that the pressure field is expressed in terms of the Jacobi determinant of the paths, that is, in terms of the local strains. Thus the leading order local gradients in the Jacobi determinant of J and Y in (9.37), and of X and J in (9.38), arise from the pressure field rather than directly from the path components X and Y. To leading order, (9.37) and (9.38) become

$$\frac{\partial^2 X}{\partial t^2} = c_s^2 J^{-\gamma-1} \left\{\beta^2 X - \alpha\beta Y\right\} + \cdots \tag{9.39}$$

$$\frac{\partial^2 Y}{\partial t^2} = c_s^2 J^{-\gamma-1} \left\{-\alpha\beta X + \alpha^2 Y\right\} + \cdots \tag{9.40}$$

where again

$$\alpha = \frac{\partial X}{\partial b}\frac{\partial}{\partial a} - \frac{\partial X}{\partial a}\frac{\partial}{\partial b}, \qquad \beta = \frac{\partial Y}{\partial b}\frac{\partial}{\partial a} - \frac{\partial Y}{\partial a}\frac{\partial}{\partial b}, \tag{9.41}$$

and the ellipses indicate space derivatives of X and Y of lower (that is, first) order. The space derivatives of X and Y in (9.15) and (9.16), which serve to characterize the well-posedness or otherwise of Lagrangian incompressible flow, are relegated to the role of coefficients in the operators α and β appearing in (9.39) and (9.40). It is readily shown that the system (9.39) and (9.40) is hyperbolic.

Exercise 9.1 The *shallow-water* equations are, in mixed form,

$$\frac{\partial^2 X}{\partial t^2} = -g\frac{\partial \chi}{\partial x}, \tag{9.42}$$

$$\frac{\partial^2 Y}{\partial t^2} = -g\frac{\partial \chi}{\partial y}, \tag{9.43}$$

$$\frac{\partial}{\partial t}(\chi J) = 0, \tag{9.44}$$

where g is the gravitational acceleration, χ is the water depth (assuming a flat bottom) and J is the Jacobi determinant

$$J = \frac{\partial(X, Y)}{\partial(a, b)}. \tag{9.45}$$

Show that the pure Lagrangian form is formally well posed. For some numerical solutions in comoving domains, see Bennett and Chua (1999). □

9.6 Labeling singularities

The preceding classification of various Lagrangian differential operators as elliptic or hyperbolic depends upon the Jacobi determinant being positive at all times. The determinant is, again,

$$J_s^t \equiv \det\left(\frac{\partial X_i}{\partial a_j}\right), \tag{9.46}$$

where $X_i = X_i(a_j, s|t)$ is the particle position at time t, and a_j is the value of a label assigned to the particle at time s. The label may be the position of the particle at time s. It is only assumed that the determinant is strictly positive at time s: $J_s^s > 0$ for all a_j. That is, the labeling is one-to-one at release. Conservation of mass (3.3) implies that

$$J_s^t = J_s^s \frac{\rho_s}{\rho}, \tag{9.47}$$

where $\rho = \rho(a_j, s|t)$ is the density and $\rho_s \equiv \rho(a_j, s|s)$. Thus a change of the sign of the determinant can only occur if the density changes sign. For isentropic flow, a change of sign of pressure is implied; see (9.33). Both density and pressure are always positive in a real fluid but the mathematical representation may fail. This may be explored further by exploiting the kinematic identity (3.20), which implies

$$J_s^t = J_s^s \exp\left(\int_s^t \Delta(r)dr\right), \tag{9.48}$$

where Δ is the flow divergence:

$$\Delta(t) \equiv \frac{\partial}{\partial x_k} u_k[x_i, t]\bigg|_{x_i = X_i(a_j, s|t)}. \tag{9.49}$$

It follows that the determinant can only change sign if Δ has a singularity which is not integrable in time. That this is a mathematical possibility may be seen by deriving the following equation for compressible flow, analagous to the nonlinear sound wave equation (3.103):

$$\frac{1}{J_s^t}\frac{\partial^2}{\partial t^2} J_s^t = \cdots, \tag{9.50}$$

where the ellipsis denotes second-order partial derivatives of J_s^t with respect to the labels a_j. A simple rearrangement of (9.48) and (9.50) yields

$$\frac{\partial \Delta}{\partial t} + \Delta^2 = \cdots, \tag{9.51}$$

where the ellipsis is as before. This equation has the potential for solution "blow up". Indeed, consider just the homgeneous Riccati equation

$$\frac{\partial \Delta}{\partial t} + \Delta^2 = 0\,. \tag{9.52}$$

The solution of (9.52) is

$$\Delta = \frac{\Delta_s}{1 + \Delta_s(t-s)}\,. \tag{9.53}$$

If $\Delta_s < 0$, the solution for Δ will have a nonintegrable singularity at time $t = s - 1/\Delta_s$, and J_s^t will change sign at that time. It is conceivable that this pathology can be suppressed by the operators (the dynamics) included in the ellipsis in (9.50), but the existence of classical solutions of the equations of gas dynamics has yet to be established. In summary, it is mathematically possible that the Lagrangian labeling may become singular in a finite time, but the implied change of sign of the thermodynamic state variables makes such a singularity physically meaningless.

9.7 Phenomenology

It is unexpected that the mathematical problem of incompressible Lagrangian flow should be ill-posed unless the pressure gradient is eliminated, while compressible Lagrangian flow is formally well-posed in either momentum or vorticity form. It is therefore appropriate to review the phenomenology of fluid dynamics. The Eulerian laws of conservation of mass, momentum and energy for compressible fluids may be derived from Boltzmann's equation for assemblies of molecules; it is necessary only to insist that molecular collisions conserve the total mass, momentum and energy involved in each collision, and that the distribution function for the molecular velocities is isotropic to first order in the Chapman–Enskog expansion (Chapman and Cowling, 1970). Indeed, for a monatomic gas without internal degrees of freedom, the ratio of specific heats γ is correctly determined to be $5/3$. Thus, compressible fluid dynamics is securely based on particle dynamics. Incompressible flow is recovered as a singular limit, in which the sound speed is infinite and the energy equation is denied a role in the dynamics. Rather, the pressure gradient is inferred as the *ad hoc* irrotational field that instantaneously ensures Eulerian solenoidality of the fluid velocity.

It remains to ask: is Eulerian incompressible flow well-posed? The answer is rigorously in the affirmative for planar flow at least (Wolibner, 1933;

Judovich, 1964; Kato, 1967). At the nonrigorous level of discussion here, the answer here is affirmative also for three-dimensional Eulerian incompressible flow. The Eulerian equations are, again:

$$\frac{\partial u_i}{\partial t} + u_j \frac{\partial u_i}{\partial x_j} = -\frac{1}{\rho}\frac{\partial p}{\partial x_i} \tag{9.54}$$

where the density ρ is a constant, and

$$\frac{\partial u_i}{\partial x_i} = 0 . \tag{9.55}$$

The mathematically more elegant approach to solving (9.54), (9.55) is to eliminate pressure by taking the curl of (9.54); invoking (9.55) yields:

$$\frac{\partial \omega_i}{\partial t} + u_j \frac{\partial \omega_i}{\partial x_j} - \omega_j \frac{\partial u_i}{\partial x_j} = 0 , \tag{9.56}$$

where the Eulerian vorticity ω_i is

$$\omega_i = \epsilon_{ijk} \frac{\partial u_k}{\partial x_j} . \tag{9.57}$$

The flow being solenoidal, assume the existence of a velocity potential A_i:

$$u_i = \epsilon_{ijk} \frac{\partial A_k}{\partial x_j} \tag{9.58}$$

which is itself solenoidal:

$$\frac{\partial A_i}{\partial x_i} = 0 , \tag{9.59}$$

and which satisfies

$$\nabla^2 A_i = -\omega_i . \tag{9.60}$$

Thus the velocity field may be determined without reference to pressure. The latter may be determined from the velocity field as the solution of the Poisson problem

$$\frac{\partial^2 p}{\partial x_k \partial x_k} = -\rho \frac{\partial u_j}{\partial x_i}\frac{\partial u_i}{\partial x_j} , \tag{9.61}$$

which obtains by taking the divergence of (9.54) and invoking (9.55). By virtue of (9.54), Neumann boundary conditions hold for p at rigid boundaries, for example. Hence it is clear that the pressure, and hence the pressure gradient in (9.54), do not depend upon local gradients of velocity, and so (9.54) may reasonably be regarded as first-order hyperbolic.[1] However, not all initial velocity fields are admissable: (9.55) must be satisfied, in particular at $t=0$.

The difficulty with Lagrangian incompressible flow is that the Lagrangian expression of the Eulerian pressure gradient involves the strain components, that is, the gradients of the path components X and Y with respect to the Lagrangian labels. The paths are the dependent variables in the momentum equations; see (9.15) and (9.16). Moreover, these strains are the leading order labeling gradients of paths appearing in these equations, and so the strains participate in the characterization of the Lagrangian equations.

The ill-posedness of the momentum equations for inviscid, incompressible flow is revealed only by the Lagrangian form but is nonetheless serious for that. It makes clear the singular nature of the high Reynolds number limit. Indeed, the singularity is a representation of the onset of turbulence in the inertial range. The Lagrangian form can be related to reality by abandoning the concept of an ideal fluid, for a real fluid.

Ill-posedness not withstanding, analytical or semi-analytical solutions of Lagrangian form exhibiting singular behavior, in a finite time in unbounded domains, have been developed for inviscid incompressible flow (Stuart, 1987), for planar inviscid and viscous incompressible flow (Childress et al., 1989) and for inviscid compressible flow (Stuart, 1998). These spatially smooth solutions are typically characterized by initial vorticity fields that are unbounded for large spatial argument.

9.8 Viscous incompressible flow

The classical well-posedness of the Eulerian equation for planar viscous incompressible flow has been established by Ladyzhenskaya (1968). An alternative and more accessible proof owes to McGrath (1968). For three-dimensional flow, classical well-posedness remains an open question (Temam, 2000). The well-posedness of the Lagrangian equations for three-dimensional viscous incompressible flow is addressed here.

[1] It is tempting to argue that (9.56) is a first-order hyperbolic equation for the vorticity ω_i; the difficulty is that the local rate-of-strain field $\partial u_i/\partial x_j$ may depend upon the local vorticity. The difficulty may be avoided by Picard iteration, as in Section (9.8.2).

9.8.1 Equations of motion

To begin, the equations of motion are restated and redeveloped. The evolution equation (5.13) for the Cauchy invariant $\boldsymbol{\varpi}$ is

$$\frac{\partial \boldsymbol{\varpi}}{\partial t} = -\nu \boldsymbol{\nabla} \times \left(\sqrt{g}^{-1} \mathsf{g} \boldsymbol{\nabla} \times \left(\sqrt{g}^{-1} \mathsf{g} \boldsymbol{\varpi} \right) \right), \tag{9.62}$$

where g is the determinant of the metric $\mathsf{g} = \mathsf{J}^T \mathsf{J}$. Equation (9.62) has two striking features: there is no "bending" of $\boldsymbol{\varpi}$ by the shear flow (that is, there is no term analagous to the fourth on the left-hand side of (3.64), but the Jacobi matrix J explicitly influences the molecular diffusion of $\boldsymbol{\varpi}$. The fluid velocity $\boldsymbol{u}$ and and hence the particle path $\boldsymbol{X}$ can be related to the Cauchy invariant $\boldsymbol{\varpi}$. Indeed, $\boldsymbol{u}$ is related to $\boldsymbol{\varpi}$ by

$$\boldsymbol{\varpi} = \boldsymbol{\nabla} \times \left(\mathsf{J}^T \boldsymbol{u} \right), \tag{9.63}$$

where the curl is taken with respect to the Lagrangian label $\boldsymbol{a}$. Next, consider the Eulerian formulation of the incompressibility condition:

$$\boldsymbol{\nabla}_x \cdot \boldsymbol{u} = \boldsymbol{0}, \tag{9.64}$$

where the subscripted divergence is taken with respect to the Eulerian position $\boldsymbol{x}$. Assuming the spatial domain to be simply connected, it follows that there is a solenoidal, vector-valued streamfunction $\boldsymbol{\Psi} = \boldsymbol{\Psi}[\boldsymbol{x}, t]$ such that

$$\boldsymbol{u} = \boldsymbol{\nabla}_x \times \boldsymbol{\Psi}, \tag{9.65}$$

where the subscripted curl is taken with respect to $\boldsymbol{x}$. Note that $\boldsymbol{\Psi}$ is only defined up to the addition of the gradient of an arbitrary scalar. The Lagrangian reformulation of (9.65) is

$$\boldsymbol{u} = \left(J_s^t \right)^{-1} \mathsf{J} \boldsymbol{\nabla} \times \left(\mathsf{J}^T \boldsymbol{\Psi} \right), \tag{9.66}$$

where J_s^t is as always the Jacobi determinant[2]. Hence $\boldsymbol{\Psi}$ is related to $\boldsymbol{\varpi}$ by

$$\boldsymbol{\nabla} \times \left(\sqrt{g}^{-1} \mathsf{g} \boldsymbol{\nabla} \times \left(\mathsf{J}^T \boldsymbol{\Psi} \right) \right) = \boldsymbol{\varpi}, \tag{9.67}$$

and again the unsubscripted curls are taken with respect to $\boldsymbol{a}$. This is an elliptic equation for $\boldsymbol{\Phi} \equiv \mathsf{J}^T \boldsymbol{\Psi}$, given $\boldsymbol{\varpi}$. The arbitrariness of $\boldsymbol{\Psi}$ permits the

[2] A true mathematician would not have to cheat, but would guess that (9.66) is the general solution of the Lagrangian continuity equation (3.67).

assumption that $\nabla \cdot \boldsymbol{\Phi} = 0$. So long as the Jacobi matrix J remains nonsingular, the vector streamfunction is then found as

$$\boldsymbol{\Psi} = (\mathsf{J}^T)^{-1} \boldsymbol{\Phi} . \tag{9.68}$$

Finally, the particle path is of course related to the fluid velocity by the linear ordinary differential equation

$$\frac{\partial \boldsymbol{X}}{\partial t} = \boldsymbol{u} . \tag{9.69}$$

9.8.2 Picard iteration

The system of equations (9.62), (9.67), (9.66) and (9.69) is, of course, nonlinear since the ubiquitous Jacobi matrix depends upon the particle paths. The nonlinear system is formally the limit of a sequence of systems of linear equations. The system is obtained by a Picard iteration on the nonlinear system; the simplest Picard iteration on (9.62) is

$$\frac{\partial \boldsymbol{\varpi}^{(n)}}{\partial t} = -\nu \nabla \times \left(\sqrt{g^{(n-1)}}^{-1} \mathsf{g}^{(n-1)} \nabla \times \left(\sqrt{g^{(n-1)}}^{-1} \mathsf{g}^{(n-1)} \boldsymbol{\varpi}^{(n)} \right) \right) , \tag{9.70}$$

where $\mathsf{g}^{(n-1)}$ is the $(n-1)^{\text{th}}$ metric. Having solved the initial value problem (9.70) for $\boldsymbol{\varpi}^{(n)}$, the nth vector streamfunction $\boldsymbol{\Psi}^{(n)}$ is found from the elliptic equation

$$\nabla \times \left(\sqrt{g^{(n-1)}}^{-1} \mathsf{g}^{(n-1)} \nabla \times \left(\left(\mathsf{J}^{(n-1)} \right)^T \boldsymbol{\Psi}^{(n)} \right) \right) = \boldsymbol{\varpi}^{(n)} , \tag{9.71}$$

where $\mathsf{J}^{(n-1)}$ is the $(n-1)^{\text{th}}$ Jacobi matrix. The nth velocity $\boldsymbol{u}^{(n)}$ is found as

$$\boldsymbol{u}^{(n)} = \left(\left(J_s^t \right)^{(n-1)} \right)^{-1} \mathsf{J}^{(n-1)} \nabla \times \left(\left(\mathsf{J}^{(n-1)} \right)^T \boldsymbol{\Psi}^{(n)} \right) . \tag{9.72}$$

It remains to relate the nth metric $\mathsf{g}^{(n)}$ to the nth Jacobi matrix $\mathsf{J}^{(n)}$. Being no more than a trivial substitution, the relation need not be linear in order for each iterated system to be readily solvable; for example

$$\mathsf{g}^{(n)} = {\mathsf{J}^{(n)}}^T \mathsf{J}^{(n)} . \tag{9.73}$$

If the sequence of Cauchy invariants $\boldsymbol{\varpi}^{(n)}$, vector streamfunctions $\boldsymbol{\Psi}^{(n)}$, velocities $\boldsymbol{u}^{(n)}$, particle paths $\boldsymbol{X}^{(n)}$, Jacobi matrices $\mathsf{J}^{(n)}$ and metrics $\mathsf{g}^{(n)}$ converges, then in the limit these fields formally satisfy the nonlinear system of equations of motion.

9.8.3 *A priori* bounds

Subject to the nonsingularity of $\mathsf{J}^{(n)}$, each set in the sequence of Picard-iterated equations defines a well-posed problem. It remains to prove that the sequence of solutions converges to a classically defined limit. This would seem simpler than the corresponding and elusive proof for the Eulerian formulation, since there should be less difficulty in establishing bounds on norms for solutions of (9.70) than for solutions of the viscous form of the Eulerian vorticity equation (3.64) (Majda and Bertozzi, 2002). To begin, consider integrals of quadratic forms for the uniterated Cauchy invariant:

Exercise 9.2 Derive from (9.62) the following time rate of change for the integral of a quadratic form for the Cauchy invariant $\boldsymbol{\varpi}$, in an unbounded labeling domain W:

$$\frac{d}{dt}\frac{1}{2}\int_W \boldsymbol{\varpi}^T \frac{\mathsf{g}}{\sqrt{g}}\boldsymbol{\varpi}\,dW = \frac{1}{2}\int_W \boldsymbol{\varpi}^T \frac{\partial}{\partial t}\left(\frac{\mathsf{g}}{\sqrt{g}}\right)\boldsymbol{\varpi}\,dW$$

$$-\nu\int_W (\boldsymbol{\nabla}\times\boldsymbol{\xi})^T \frac{\mathsf{g}}{\sqrt{g}}\boldsymbol{\nabla}\times\boldsymbol{\xi}\,dW\,, \tag{9.74}$$

where $\boldsymbol{\xi}\equiv(\mathsf{g}/\sqrt{g})\boldsymbol{\varpi}$. Verify that the integral on the left-hand side of (9.74) is simply the total squared vorticity, and that the first integral on the right-hand side is a consequence of the bending term in (3.64). That is, (9.74) is no more than the Lagrangian formulation of the conventional budget for total squared vorticity in an unbounded domain V^t. In particular, the integrated quadratic form for the Cauchy invariant is *not* conserved, even in inviscid flow ($\nu=0$). Thus the fundamental difficulty, encountered when deriving *a priori* vorticity estimates in the Eulerian formulation, arises again in the Lagrangian formulation even though the Cauchy invariant *is* conserved by inviscid flow; see (3.50). □

9.8.4 The viscous operator

In spite of the influence of vortex bending on squared integral bounds for the Cauchy invariant $\boldsymbol{\varpi}$, the Lagrangian growth rate of the Cauchy invariant would appear, according to (9.62), to scale with the kinematic viscosity ν. The unusual nature of the evolution equation (9.62) is revealed by expressing its right-hand side as

$$\frac{\partial\boldsymbol{\varpi}}{\partial t} = \mathsf{A}\boldsymbol{\varpi} + \mathsf{B}^l\frac{\partial\boldsymbol{\varpi}}{\partial a_l} + \mathsf{C}^{lm}\frac{\partial^2\boldsymbol{\varpi}}{\partial a_l \partial a_m}\,, \tag{9.75}$$

where

$$A_{ij} = -\nu\epsilon_{ilk}\frac{\partial}{\partial a_l}\left(h_{kp}\epsilon_{pmn}\frac{\partial h_{nj}}{\partial a_m}\right), \tag{9.76}$$

$$B^l_{ij} = -\nu\left(\epsilon_{ilk}h_{kp}\epsilon_{pmn}\frac{\partial h_{nj}}{\partial a_m} + \epsilon_{imk}\frac{\partial}{\partial a_m}\left(h_{kp}\epsilon_{pln}h_{nj}\right)\right), \tag{9.77}$$

$$C^{lm}_{ij} = -\nu\epsilon_{ilk}h_{kp}\epsilon_{pmn}h_{nj}, \tag{9.78}$$

and for clarity, $h_{nm} = g_{nm}/\sqrt{g}$.

The three operators on the right-hand side of (9.75) will be examined in turn. In general, none of the three coefficient matrices is symmetric, much less positive definite. Accordingly their eigenvalues may be positive, zero, negative or complex conjugate pairs.

(i) The first operator, a matrix multiplication by A, could yield local growth, stagnation, or decay of the Cauchy invariant, all possibly combined with rotation. The inverse time-scale is ν/λ^2, where λ is the length-scale of variation of the Jacobi matrix.

(ii) Let k_l be a wavenumber vector for $\boldsymbol{\varpi}$. If the eigenvalues of $k_l\mathsf{B}^l$ are real then the second operator yields a drift of the field $\boldsymbol{\varpi}$ at a speed having the scale ν/λ. If the eigenvalues are complex and have a positive real part, then there is "Hadamard fluting" of the field with growth rate $\nu k/\lambda$, where k is the wavenumber magnitude or inverse length-scale of $\boldsymbol{\varpi}$.

(iii) The third operator could yield diffusive decay, stagnation or antidiffusive growth (another form of Hadamard fluting), all possibly combined with skew-diffusive rotation, and all having inverse time-scale νk^2. Diffusive decay, corresponding to all eigenvalues of $\mathsf{D} = -k_l\mathsf{C}^{lm}k_m$ having negative real parts, is in general essential if the forward problem for (9.62) is to be well-posed. Such decay will dominate any growth arising from the lower order operators

Exercise 9.3 Show that the eigenvalues of D are 0, $-\mu$ and $-\mu$, where

$$\mu = \nu\boldsymbol{k}^T\mathsf{g}^{-1}\boldsymbol{k}. \tag{9.79}$$

Show also that the eigenvector associated with the null eigenvalue is $\sqrt{g}\mathsf{g}^{-1}\boldsymbol{k}$. □

The null eigenvector for D does not nullify the wavenumber space representations of the lower order operators in (9.75), but the Cauchy invariant field $\boldsymbol{\varpi} = \sqrt{g}\mathsf{g}^{-1}\boldsymbol{\nabla}\phi$ nullifies the right-hand side of (9.62), for any scalar field $\phi(a_i, s|t)$. The corresponding vorticity is irrotational: $\boldsymbol{\omega} = \boldsymbol{\nabla}_x \phi$, where the gradient is taken with respect to the Eulerian variable x_i. In summary, the solutions of (9.62) either stagnate or decay in time. That is, in the absence of external forcing, the Cauchy invariant is bounded above by its initial values.

Assume now that the initial Cauchy invariant $\boldsymbol{\varpi}_s = \boldsymbol{\varpi}(a_i, s|s)$ has Hölder-continuous second spatial derivatives (Adams, 1975), that is, $\boldsymbol{\varpi}_s \in C^{2,\alpha}$ for some $\alpha > 0$. Suppose also that each coefficient matrix in (9.75) is Hölder continuous in space with index α (it suffices that $\mathsf{J} \in C^{2,\alpha}$). Then (Il'in, Kalashnikov and Oleinik, 1962; Friedman, 1964 or Bennett and Kloeden, 1981) there is a unique solution of (9.75) which also has Hölder-continuous second derivatives in space for all $t > s$: $\boldsymbol{\varpi} \in C^{2,\alpha}$. Hölder continuity of the coefficient matrices in (9.75) with respect to time carries over to the solution; time is not tricky here and will not be discussed further. Note also that an estimate for the $C^{2,\alpha}$ norm of $\boldsymbol{\varpi}$ will depend upon the initial value $\boldsymbol{\varpi}_s$ and the coefficient matrices $\mathsf{A}, \mathsf{B}, \mathsf{C}$.

9.8.5 The elliptic operator

Assuming that an *a priori* estimate has been obtained for the Cauchy invariant $\boldsymbol{\varpi}$, an estimate must then be obtained for the particle paths $\boldsymbol{X}$ via the velocity field $\boldsymbol{u}$, which is constructed from the strained vector streamfunction $\boldsymbol{\Phi}$ using (9.66) and (9.68), while $\boldsymbol{\Phi}$ is related to $\boldsymbol{\varpi}$ through (9.67). The last mentioned equation is understood to have been subjected to a Picard iteration, which yields a linear elliptic equation with coefficients evaluated at the preceding iterate; see (9.71). The iteration indices are suppressed here for clarity. Estimates for the solution of this equation will involve estimates for first spatial derivatives of the metric g, that is, second derivatives of the particle path $\boldsymbol{X}$. At this point it would seem advantageous to cheat.

The Eulerian field of the Cauchy invariant is

$$\boldsymbol{\varpi}[\boldsymbol{x}, t] = \boldsymbol{\varpi}(\boldsymbol{A}(\boldsymbol{x}, t|s), s|t)\,. \tag{9.80}$$

The "inverted" path $\boldsymbol{A}(\boldsymbol{x}, t|s)$ in (9.80) is assumed known from the preceding iterate. The Jacobi matrix J and determinant J_s^t may similarly be converted to Eulerian fields, and hence also the vorticity $\boldsymbol{\omega}$:

$$\boldsymbol{\omega} = (J_s^t)^{-1} \mathsf{J} \boldsymbol{\varpi} \tag{9.81}$$

(see (3.69)). The solenoidal Eulerian vector streamfunction $\boldsymbol{\Psi}$ (see (9.65)) is related to the vorticity via the Poisson problem

$$\nabla_x^2 \boldsymbol{\Psi} = -\boldsymbol{\omega}\,, \tag{9.82}$$

for which estimates and existence theorems abound (see, for example, Theorem 6.8 of Gilbarg and Trudinger, 1983). These estimates for second derivatives of $\boldsymbol{\Psi}$ depend upon upon at least, for example, an undifferentiated Hölder estimate (Adams, 1975) for $\boldsymbol{\omega}$ as an Eulerian field. It is clear from (9.81) that the same estimate is required of $\boldsymbol{\varpi}$ and of J as Eulerian fields. Examination of (9.80) shows that these in turn require the same estimates as Lagrangian fields, and that the same estimate is also required for the particle path $\boldsymbol{X}$. The last mentioned estimate is assured by the estimate for J. In summary, the elliptic equation is vastly simplified by cheating, and the required estimate for $\boldsymbol{X}$ is lower: a once-differentiated Hölder estimate for the particle paths is required, rather than twice differentiated as required by (9.67). The function spaces are:

$$\boldsymbol{\varpi}, \mathsf{J} \in C^{\alpha}, \quad \boldsymbol{\Psi} \in C^{2,\alpha}, \quad \boldsymbol{u}, \boldsymbol{X} \in C^{1,\alpha}, \tag{9.83}$$

for some $\alpha > 0$. Yet the coefficient matrices in (9.75) must be in C^{α}, thus at the very least J must be in $C^{2,\alpha}$. Happily, the chain

$$\boldsymbol{\varpi}, \mathsf{J} \in C^{m,\alpha}, \quad \boldsymbol{\Psi} \in C^{m+2,\alpha}, \quad \boldsymbol{u}, \boldsymbol{X} \in C^{m+1,\alpha}, \tag{9.84}$$

$m \geq 2$, is consistent with (9.62), (9.81), (9.82), (9.65) and (9.69).

Exercise 9.4 Why not reduce the number of space derivatives of g appearing in (9.62) by substituting for $\sqrt{g}^{-1}\mathsf{g}\boldsymbol{\varpi}$? □

Exercise 9.5 Prove that the velocity sequence $\boldsymbol{u}^{(n)}$ contructed in Section (9.8.2) converges to a classical solution of the Navier–Stokes equations in any finite time interval $s < t < t_1$. □

The preceeding discussion tacitly assumes a pure initial value problem. That is, conditions at finitely distant boundaries have not been considered. There is a difficulty in determing values for the Cauchy invariant at a no-slip boundary, just as there is for the Eulerian vorticity; see e.g. Rosenhead (1963), Foreman and Bennett (1988).

PART III

Diffusion

Introduction

A principal objective of any theory of fluid motion is the prediction of the spread of matter or "tracer" within the fluid. The problem is trivial for the fluid particles themselves in steady flow: they follow streamlines. It is nontrivial if the motion is time dependent, or if the tracer is dissolved in the fluid but diffusing through it. The time dependence of general interest is turbulence. The next four chapters develop a coherent framework for considering inhomogeneous and nonstationary turbulence, with elaboration in detail for the homogeneous, stationary and incompressible case, excluding and including tracer diffusion. Applications to the spread of phytoplankton are of special interest to oceanographers; these marine organisms are modeled as reacting tracers having nonlinear reaction rates.

Absolute dispersion is considered first. This is the problem of predicting the path of a single fluid particle, or the path of the centroid of a cluster of particles, in turbulent flow. Turbulence being conceived as a random process, the problem is the prediction of the probability distribution function or pdf for the particle path. The mathematical difficulty is the closure of the infinite heirarchy of moments of the nonlinear kinematics, that is, the relating of certain high-order moments of particle displacement to low-order moments. There are any number of workings of this task in the literature, most of which close at second order, that is, second moments are related to first. All workings invoke the rapidity of loss of correlation along the particle path, in some fashion or another. The formulation presented here is about as simple as can be. Regardless, closure is an intricate business and questions of reversibility must be addressed. Incompressibility of the flow ensures reversibility, leading immediately to what is known as the semi-empirical equation of turbulent diffusion. The equation is parabolic; the diffusivity is in general spatially nonuniform. Higher order closures may be constructed but they can be highly misleading since, in the special case of homogeneous, stationary, incompressible flow with exactly normally distributed particle paths, a perfect closure is available and the result is a special case of second-order closure. The evolution equations arising in second-order closures are identifiable as Fokker–Planck equations for Markov processes; this formally expresses the assumption of rapid decorrelation along the path. Numerical integration of the stochastic differential equations for the Markov processes themselves is an appealing technique for solving the semi-empirical equation but no more theoretical hydrodynamics is being invoked, and that which is invoked – the second-order closure – may be wrong. Much of the theory of absolute dispersion consists of arguments for particular forms of the diffusivity in the semi-empirical equation (see, for example, Monin and Yaglom, 1971). Yet the whole form of the closure may be wrong. Nevertheless, armed with this closure, much has been computed. The perfect closure, for absolute dispersion of phytoplanton populations living and dying in an annual cycle, produces spatial patchiness in the populations, even if they are adrift in that maximally featureless ocean.

Relative dispersion describes the spreading of an initially close cluster of fluid particles. Their motions are correlated whenever they are close, which is an early probability. The especially simple closure for single particles is extended to a pair in isotropic and incompressible turbulence; the result is the remarkable equation of relative dispersion *guessed* by Richardson (1926), and deduced by Kraichnan (1966b) with an elaborate closure. The equation is again parabolic, but the diffusivity is a function of particle separation. The form of this "relative" diffusivity may be deduced by dimensional arguments for several classes of isotropic turbulence (now understood also to be stationary and incompressible). These forms are consistent with at least some relevant atmospheric and oceanic data.

The variance spectrum for nondiffusing tracers in isotropic turbulence may be derived from the separation pdf. Several wavenumber regimes exist, according to the significance of viscosity in the dynamics of the turbulence. The variance of a phytoplankton population, which population has a spatially featureless growth rate and is dispersing in isotropic turbulence, is unable to retain any initial patchiness.

The Lagrangian conservation law for a diffusing tracer has an elementary solution, if the spatial dependencies of the Jacobi matrix elements are ignored. The simplification turns out to be justified for an interesting class of parameter regimes, thus a unified theory of the diffusive subranges of the scalar variance spectrum may be presented. Some of the universal spectral forms predicted by the theory (and also by other more arcane analyses) have been verified experimentally; one may be impossible to verify.

Other predictions are spectral forms which are dependent upon the shape of the variance spectrum for the tracer source, that is, the predicted forms are not universal. Various regimes are identifiable in flows as disparate as microstructure in liquid mercury, deep ocean currents, and circulation in the stratosphere.

10
Absolute dispersion

10.1 Displacement: first and second moments

For simplicity consider rectilinear motion first. The Lagrangian velocity, or velocity at time t of a particle having path $X = X(a, s|t)$, satisfies

$$\frac{\partial}{\partial t} X(a, s|t) = u(a, s|t), \quad X(a, s|s) = a. \tag{10.1}$$

Note the specific choice of particle label, namely, the particle position a at time s. The solution of (10.1) may be expressed as

$$X(a, s|t) = a + \int_s^t u(a, s|r) dr. \tag{10.2}$$

Imagine that the Lagrangian velocity is random; that is, imagine it may assume any one of an ensemble of values such that the integral in (10.2) has meaning in some reasonable sense. Then the expectation or ensemble mean of the particle displacement $X(a, s|t) - a$ is

$$E\{X(a, s|t) - a\} = \int_s^t E\{u(a, s|r)\} dr. \tag{10.3}$$

Note that the expectation is *conditioned* by the position a and time s of release. That is, the expected displacement is a Lagrangian mean. In particular, the expectation of the Lagrangian velocity in (10.3) is the mean of all velocities at time r experienced by particles released at (a, s).

The Lagrangian variance of the displacement is

$$\mathrm{var}\{X(a, s|t)\} = E\left\{\left(X(a, s|t) - E\{X(a, s|t)\}\right)^2\right\}, \tag{10.4}$$

which satisfies

$$\frac{\partial}{\partial t} \mathrm{var}\{X(a, s|t)\} = 2\kappa(a, s|t), \tag{10.5}$$

where, defining $u(a, s|t)' \equiv u(a, s|t) - E\{u(a, s|t)\}$,

$$\kappa(a, s|t) = \int_s^t E\{u(a, s|t)'u(a, s|r)'\}dr \tag{10.6}$$

is the *Taylor diffusivity* (Taylor, 1921). Note that $X(a, s|s) = a$, so $\text{var}\{X(a, s|s)\} = 0$. In the simplest case of Lagrangian homogeneous and Lagrangian stationary turbulence, $E\{u(a, s|t)\} = U$ and $E\{u(a, s|t)' u(a, s|r)'\} = \mu^2 C(t-r)$, where U and μ are constants, while $C = C(t)$ is the velocity decorrelation function. If C is integrable:

$$\int_0^\infty C(t)dt = \tau < \infty, \tag{10.7}$$

then τ is the "Lagrangian integral time scale" and

$$\text{var}\{X(a, s|t)\} \sim 2\kappa_\infty(t-s) \tag{10.8}$$

as $t \to \infty$, where $\kappa_\infty = \mu^2\tau$ is the Taylor diffusivity constant. The asymptotic behavior (10.8) is characteristic of a *random walk* (e.g., Gardner, 1985).

Generalization to two or three dimensions is straightforward. Denoting the particle path by $X_k(a_j, s|t)$, the ensemble mean displacement vector is

$$E\{X_k(a_j, s|t) - a_k\} = \int_s^t E\{u_k(a_j, s|r)\}dr, \tag{10.9}$$

while the *single-particle, single-time* displacement covariance tensor $\text{cov}\{X_i, X_j\}$ satisfies

$$\begin{aligned}\frac{\partial}{\partial t}\text{cov}\{X_i(a_k, s|t), X_j(a_k, s|t)\} &\equiv \frac{\partial}{\partial t} E\{X_i(a_k, s|t)'X_j(a_k, s|t)'\} \\ &= \kappa_{ij}(a_k, s|t) + \kappa_{ji}(a_k, s|t).\end{aligned} \tag{10.10}$$

There is no summation over the repeated index k on the labeling arguments in (10.10), *et seq*. The Taylor diffusivity tensor in (10.10) is

$$\kappa_{ij}(a_k, s|t) = \int_s^t \text{cov}\{u_i(a_k, s|t), u_j(a_k, s|r)\}dr. \tag{10.11}$$

Note that κ_{ij} is not symmetric, but only its symmetric part influences the variance of displacement (10.10).

10.2 Displacement pdf

Consider rectilinear motion first. The integral representation of displacement (10.2) may be approximated by a Riemann sum:

$$X_M = a + \Delta t \sum_{m=0}^{M-1} u_m , \tag{10.12}$$

where $u_m = u(a, s|t_m)$, $X_M = X(a, s|t_M)$ and $t_m = m\Delta t + s$. Assume that the turbulent flow is statistically stationary in the Lagrangian sense. That is, along the path,

$$E\{u_m\} = U, \ \text{var}\{u_m\} = \mu^2 \tag{10.13}$$

where U, μ are constants, for fixed (a, s). Simpler still, assume u_n to be uncorrelated if $n \neq m : \text{cov}\{u_n, u_m\} = \mu^2 \delta_{nm}$. The central limit theorem (e.g., Gnedenko, 1976) establishes that

$$X_M \sim \mathcal{N}\big(E\{X_M\}, \ \text{var}\{X_M\}\big) . \tag{10.14}$$

as $M \to \infty$. That is, X_M is asymptotically normally distributed for large M. The mean and variance are:

$$E\{X_M\} = a + U\big(t_M - s\big) , \tag{10.15}$$

$$\text{var}\{X_M\} = 2K\big(t_M - s\big) . \tag{10.16}$$

The constant diffusivity in (10.16) is $K = \frac{1}{2}\Delta t \mu^2$. In other words, the probability distribution function for displacement satisfies

$$P(X_M) \sim \big(2\pi \text{var}\{X_M\}\big)^{-\frac{1}{2}} \exp\left[-\frac{\big(X_M - E\{X_M\}\big)^2}{2\text{var}\{X_M\}} \right] , \tag{10.17}$$

as $M \to \infty$. In the case of nonstationary but uncorrelated Lagrangian velocities, that is,

$$E\{u_m\} = U_m , \ \text{cov}\{u_n, u_m\} = \mu_m^2 \delta_{nm} , \tag{10.18}$$

the asymptotic result (10.14) still holds, now with

$$E\{X_M\} = a + \overline{U}_M \big(t_M - s\big) , \tag{10.19}$$

$$\text{var}\{X_M\} = 2\overline{K}_M \big(t_M - s\big) , \tag{10.20}$$

where $\overline{K}_M = \frac{1}{2}\Delta t \left(\overline{\mu^2}\right)_M$, while the overbars denote arithmetic means:

$$\overline{U}_M = M^{-1} \sum_{m=0}^{M-1} U_m , \quad \left(\overline{\mu^2}\right)_M = M^{-1} \sum_{m=0}^{M-1} \mu_m^2 . \tag{10.21}$$

Proof of the central limit theorem (CLT) in this nonstationary case requires satisfaction of the Lindeberg condition (Gnedenko, 1976). Roughly speaking, the condition is that the total variance in the sum must not be dominated by a finite number of terms. That is, the number of degrees of freedom must grow without bound for large M. The proof in the nonstationary case may be generalized (Cocke, 1972) to admit correlated Lagrangian velocities:

$$\text{cov}\{u_n, u_m\} \neq 0, \text{ for } n \neq m. \tag{10.22}$$

There will be fewer degrees of freedom than in the uncorrelated case but, if the CLT is to hold, the number must still grow without bound for large M. The most inclusive and rigorous expressions of this condition are so subtle that it is unlikely they could be tested experimentally, nor is it likely they could be deduced from the Navier –Stokes equation.

Let us now assume that while the Lagrangian velocity may be nonstationary and correlated, the resulting *continuous* displacement (10.2) is asymptotically normal long after release. That is,

$$X(a_i, s|t) \sim \mathcal{N}\Big(E\big\{X(a, s|t)\big\}, \text{var}\big\{X(a, s|t)\big\}\Big) \tag{10.23}$$

as $t \to \infty$, where the mean and variance are given by (10.3) and (10.5). So $P(a, s|X, t)$, the large-time asymptotic probability distribution function (pdf) for $X(a, s|t)$, is given by

$$P(a, s|X, t) \sim \big(2\pi \text{var}\{X\}\big)^{-\frac{1}{2}} \exp\left[-\frac{\big(X - E\{X\}\big)^2}{2\text{var}\{X\}}\right], \tag{10.24}$$

as $t \to \infty$. It is easy to show that (10.24) satisfies

$$\frac{\partial}{\partial t} P(a, s|X, t) = -E\big\{u(a, s|t)\big\} \frac{\partial}{\partial X} P(a, s|X, t) + \kappa(a, s|t) \frac{\partial^2}{\partial X^2} P(a, s|X, t), \tag{10.25}$$

as $t \to \infty$. Note that (10.25) is is said to be a linear diffusion equation, since the coefficients are independent of X and so the time rates of change of the first and second moments of X are independent of X: see (10.3) and (10.5). The initial condition for (10.25) is

$$P(a, s|X, s) = \delta(X - a), \tag{10.26}$$

which is no more than restating that $X(a, s|s) = a$ for all paths in the ensemble.

In the multidimensional case, the CLT yields

$$X_k(a_i, s|t) \sim \mathcal{N}\big(E\{X_k\}, \text{cov}\{X_j, X_l\}\big) \tag{10.27}$$

as $t \to \infty$. That is,

$$P(a_j, s|X_k, t) \sim (2\pi)^{-\frac{D}{2}} S^{-\frac{1}{2}} \exp\left[-\frac{1}{2}Q\right] \tag{10.28}$$

where $D=2$ or 3 is the spatial dimension, while

$$S = \det\left(\text{cov}\{X_j, X_k\}\right), \quad Q = \left(X_j - E\{X_j\}\right) w_{jk} \left(X_k - E\{X_k\}\right). \tag{10.29}$$

In (10.29), w_{jk} is an element of the inverse of the matrix having elements $\text{cov}\{X_j, X_k\}$. It is a tedious exercise to show that (10.29) satisfies

$$\frac{\partial}{\partial t} P(a_j, s|X_k, t) = -E\left\{u_l(a_j, s|t)\right\} \frac{\partial}{\partial X_l} P(a_j, s|X_k, t) + \frac{1}{2}\left(\kappa_{nm}(a_j, s|t) + \kappa_{mn}(a_j, s|t)\right) \frac{\partial^2}{\partial X_n \partial X_m} P(a_j, s|X_k, t) \tag{10.30}$$

where the coefficients are, respectively, the mean Lagrangian velocity and the symmetric part of the Taylor diffusivity (10.11). The linear diffusion equation (10.30) is subject to the initial condition

$$P(a_j, s|X_k, s) = \delta(X_j - a_j). \tag{10.31}$$

Note that the right-hand side (rhs) of (10.31) is the product of two or three Dirac delta functions, according to the dimensionality of X_j. The explicit formula (10.28) and the linear diffusion equation (10.30), (10.31) contain the same information about the large-time asymptotic pdf for X_k. Neither is of value without a knowledge of $E\{X_k\}$ and $\text{cov}\{X_j, X_l\}$. Both the formula and the linear diffusion equation express the CLT, the proof of which requires that the domain for X_k be unbounded in all directions.

10.3 Forward closure, boundary conditions

The CLT being of no avail in partially or totally bounded domains, recourse is made to closure theory in order to devise an equation for the approximation evolution of the pdf for X_k. To begin, choose one member of the ensemble of time series of Lagrangian velocities all labeled by (a_k, s). Then, without doubt the so-called "micro" pdf for this one member is

$$\pi(a_k, s|x_j, t) = \delta\left(x_j - X_j(a_k, s|t)\right). \tag{10.32}$$

That is, x_j definitely coincides with the position of the particle at time t, provided $x_j = X_j(a_k, s|t)$. The "macro" pdf for the ensemble is then

$$P(a_k, s|x_j, t) = E\{\pi(a_k, s|x_j, t)\}. \tag{10.33}$$

Aside: The notation on the left-hand side (lhs) of (10.33) is inferior but conventional. A superior notation (Feller, 1968) would be

$$P_{X_j}(a_k, s|x_j, t) = E\{\pi(a_k, s|x_j, t)\}$$

where the lhs denotes the probability distribution function for the random variable X_j, conditioned by a_k and s, taking values near x_j at time t. The inferior notation (10.33) will be retained in the interest of simplicity, but the reader should occasionally review the meaning of the all-purpose symbol P.

It follows easily from (10.32) and (10.33) that

$$\frac{\partial}{\partial t}P(a_i, s|x_j, t) = -E\{u_k(a_i, s|t)\}\frac{\partial}{\partial x_k}P(a_i, s|x_j, t)$$
$$-E\left\{u_k(a_i, s|t)'\frac{\partial}{\partial x_k}\pi(a_i, s|x_j, t)'\right\}, \tag{10.34}$$

where $\pi' = \pi - E\{\pi\}$ is the fluctuation of the micro pdf about the macro pdf. Note that the coefficients $E\{u_k\}$ and $u_k{}'$ in (10.34) have as arguments $(a_i, s|t)$, but *not* x_j. This equation for P becomes *closed* by expressing the covariance term (the second term) on the rhs of (10.34) solely with *moments* of u_k and operators acting on P.

To this end, the following equation for $\pi(a_i, s|x_j, t)'$ is readily derived:

$$\frac{\partial \pi'}{\partial t} + u_k\frac{\partial \pi'}{\partial x_k} = T_1 + T_2, \tag{10.35}$$

where

$$T_1(a_i, s|x_j, t) = E\left\{u_k(a_i, s|t)'\frac{\partial}{\partial x_k}\pi(a_i, s|x_j, t)'\right\}, \tag{10.36}$$

$$T_2(a_i, s|x_j, t) = -u_k(a_i, s|t)'\frac{\partial}{\partial x_k}P(a_i, s|x_j, t). \tag{10.37}$$

Note that the advecting velocity on the lhs of (10.35) is the total Lagrangian velocity $u_k(a_i, s|t)$, and note that T_1 is deterministic.

The initial condition for π' is

$$\pi(a_i, s|x_j, s)' = 0. \tag{10.38}$$

The solution of (10.35)–(10.38) is easily seen to be

$$\pi(a_i, s|x_j, t)' = \sum_{n=1}^{2} \int_s^t T_n\left(a_i, s|x_j - \int_r^t u_j(a_i, s|w)dw, r\right) dr, \tag{10.39}$$

which may be used to express the covariance in (10.34). So far this analysis is exact. Two approximations will now be made.

(i) The random Lagrangian velocity u_j appearing in the argument of T_1 in (10.39) is replaced with its mean $E\{u_j\}$. Consequently, the contribution involving T_1 to the covariance in (10.34) is proportional to $E\{u_k{}'\}$, which vanishes. This approximation will be discussed in Section 10.7.
(ii) The term T_2 contributes the covariance

$$E\left\{u_k(a_i, s|t)' \frac{\partial}{\partial x_k} \pi(a_i, s|x_j, t)'\right\} = -\int_s^t E\left\{u_k(a_i, s|t)' u_l(a_i, s|r)' \frac{\partial^2}{\partial x_k \partial x_l} P\left(a_i, s|x_j - \int_r^t u_j(a_i, s|w)dw, r\right)\right\} dr. \tag{10.40}$$

There is a standard quasi-Markovian approximation (see, for example, Lundgren (1981), Drummond (1982) and their references), which must inevitably be made in order to obtain a *second-order* closure, that is, a closure involving only second moments of the Lagrangian velocity field u_k. In applying the approximation to (10.40), the assumed decorrelation of the Lagrangian velocities for large $t-r$ is employed as a crude justification for replacing r by t in the arguments of P. This too will be discussed in Section 10.7. Substituting the resulting approximation into (10.34) yields the following evolution equation for $P(a_k, s|x_j, t)$:

$$\frac{\partial}{\partial t} P(a_i, s|x_j, t) = -E\{u_k(a_i, s|t)\} \frac{\partial}{\partial x_k} P(a_i, s|x_j, t) + \kappa_{nm}(a_i, s|t) \frac{\partial^2}{\partial x_n \partial x_m} P(a_i, s|x_j, t). \tag{10.41}$$

This is identical to the linear diffusion equation (10.30), since the mixed second derivatives of P are assumed to be symmetric. The initial condition is again (10.31).

Assuming an unbounded domain, the solution of (10.41), (10.31) is multivariate normal with moments satisfying

$$\frac{\partial}{\partial t} E\{x_k\} = E\{u_k\}, \quad \frac{\partial}{\partial t} \mathrm{cov}\{x_j, x_l\} = \kappa_{jl} + \kappa_{lj}. \tag{10.42}$$

These are no more than identities. They cannot be satisfied, however, by solutions of (10.41) in partially or completely bounded domains. Indeed, if the domain is rigidly bounded, so that a particle cannot escape, then normalization of P requires that

$$\int PdV(x) = 1\,, \tag{10.43}$$

where the integral is over the domain of x_k and the volume element is $dV(x)$. This is satisfied by the initial condition (10.31). Integrating (10.41) yields

$$\frac{\partial}{\partial t}\int PdV(x) = -E\{u_k\}\oint \hat{n}_k PdA(x) + \kappa^S_{kj}\oint \hat{n}_j \frac{\partial P}{\partial x_k} dA(x)\,, \tag{10.44}$$

where $\hat{n}_k$ is the unit outward normal on the boundary, and the boundary integrals have area element $dA(x)$. Thus it suffices that

$$-E\{u_k(a_i, s|t)\}\,\hat{n}_k(\xi_l)P(a_i, s|\xi_l, t) + \kappa^S_{kj}(a_i, s|t)\hat{n}_j(\xi_l)\frac{\partial}{\partial \xi_k}P(a_i, s|\xi_l, t) = 0 \tag{10.45}$$

at each boundary point ξ_l, where κ^S_{kj} is the symmetric part of the Taylor diffusivity tensor. That is, the inward advective flux of probability across the boundary must balance the outward diffusive flux. Again, note that the Lagrangian mean velocity and Taylor diffusivity are independent of ξ_l.

The first and second moments of (10.41) with respect to x_k yield

$$\frac{\partial}{\partial t}E\{x_k\} = E\{u_k\} - \kappa^S_{ik}\oint \hat{n}_i PdA(x)\,, \tag{10.46}$$

$$\frac{\partial}{\partial t}\mathrm{cov}\{x_k, x_l\} = 2\kappa^S_{kl} - \kappa^S_{ik}\oint \hat{n}_i x_l PdA(x) - \kappa^S_{il}\oint \hat{n}_i x_k PdA(x) \tag{10.47}$$

after invoking (10.45). These are correct only if the boundary integrals vanish. If the boundary is rigid, then any particle released inside the domain can never reach the boundary and so the micro distribution $\pi(a_k, s|x_j, t)$ must vanish on the boundary. Hence, the macro distribution must also vanish:

$$P(a_k, s|x_j, t) = 0\,, \tag{10.48}$$

for a_k within the domain, and x_j on the boundary. Imposing this consequence of boundary rigidity would annihilate the boundary integrals in (10.46) and (10.47). Yet it is not permissable to impose both (10.45) and (10.48) upon the diffusion equation (10.41). It may be remarked that there is no problem with moments of the exact equation (10.34), since π' vanishes on the boundary. This freedom to specify boundary conditions on P and π' exists since (10.34), unlike (10.41), is not closed. The dangers of approximate closure are evident.

10.4 Backward closure, scalar concentrations

The labeling theorem for the micro pdf $\pi(a_l, s|x_j, t)$ is

$$\frac{\partial}{\partial s}\pi(a_l, s|x_j, t)+u_k[a_l, s]\frac{\partial}{\partial a_k}\pi(a_l, s|x_j, t)=0, \tag{10.49}$$

where $u_k[a_l, s]=u_k(a_l, s|s)$ is the Eulerian velocity at release. The final condition is

$$\pi(a_l, t|x_j, t)=\delta(a_l-x_l). \tag{10.50}$$

The simple, approximate closure of Section 10.3 may be applied to (10.49), (10.50), yielding the backward diffusion equation

$$\frac{\partial}{\partial s}P(a_l, s|x_j, t)+E\{u_k[a_l, s]\}\frac{\partial}{\partial a_k}P(a_l, s|x_j, t)$$
$$=-\int_s^t E\left\{u_k[a_l, s]'\frac{\partial}{\partial a_k}\left(u_m(a_l, s|r)''\frac{\partial}{\partial a_m}P(a_l, s|x_j, t)\right)\right\}dr. \tag{10.51}$$

Notes:

(i) the case of interest is $s<t$;
(ii) the gradient of P in (10.51) is both deterministic and independent of the time integration variable r;
(iii) the double prime indicates a mixed Eulerian–Lagrangian fluctuation:

$$u_m(a_l, s|r)''=u_m[\xi_k, r]', \tag{10.52}$$

evaluated at $\xi_k=X_k(a_l, s|r)$;
(iv) the coefficients in (10.51) depend upon the independent spatial variable a_j.

That (10.51) is a diffusion equation will be clearer when incompressible flow is considered shortly. But first consider the utility of the backward equation.

Let C be the concentration of a tracer per unit mass of fluid, that is, a scalar obeying the Lagrangian conservation law

$$\frac{\partial}{\partial t}(\rho J_s^t C)=\rho J_s^t S, \tag{10.53}$$

where ρ is the density, J_s^t is the Jacobi determinant of the particle paths and $S=S(a_i, s|t)$ is the tracer source strength along the path. The initial condition is

$$C(a_i, s|s)=C_I[a_i, s]. \tag{10.54}$$

After invoking the Lagrangian conservation law for mass, it is readily verified that

$$C[x_i, t] = C_I[X_j(x_i, t|s), s] + \int_s^t S[X_j(x_i, t|r), r]dr. \tag{10.55}$$

Expressing (10.55) as

$$C[x_i, t] = \int C_I[\xi_j, s]\delta(\xi_j - X_j(x_i, t|s))dV(\xi) + \int_s^t \int S[\xi_j, r]\delta(\xi_j - X_j(x_i, t|r))dV(\xi)dr, \tag{10.56}$$

where $dV(\xi)$ denotes the volume element in ξ_i space. Taking averages for the ensemble of paths passing through (x_i, t) yields

$$E\{C[x_i, t]\} = \int C_I[\xi_j, s]P(x_i, t|\xi_j, s)dV(\xi) + \int_s^t \int S[\xi_j, r]P(x_i, t|\xi_j, r)dV(\xi)dr. \tag{10.57}$$

Notes

(i) It is assumed that the paths are statistically independent of both the initial scalar concentration C_I and the source strength S. That is, some dynamical influence is responsible for the randomness of the paths. The scalar is "passive" in this statistical sense.

(ii) The Eulerian arguments (x_i, t) are the Lagrangian labels for P in the integrals on the rhs of (10.57), while the integration variables (ξ_j, r) are parameters in (10.51), so

$$\frac{\partial}{\partial t}E\{C[x_i, t]\} + E\{u_k[x_i, t]\}\frac{\partial}{\partial x_k}E\{C[x_i, t]\} = \int_s^t E\left\{u_k[x_i, t]'\frac{\partial}{\partial x_k}\left(u_l(x_i, t|r)''\frac{\partial}{\partial x_l}E\{C[x_i, t]\}\right)\right\}dr + S[x_i, t]. \tag{10.58}$$

This is almost a purely Eulerian equation, corrupted only by the mixed Eulerian–Lagrangian nature of the diffusion coefficient. The terminals in the integral are correct!

Consider the case of incompressible flow:

$$\frac{\partial}{\partial x_k}u_k[x_i, t] = 0. \tag{10.59}$$

Then (10.58) simplifies to

$$\frac{\partial}{\partial t}E\{C[x_i,t]\}+E\{u_k[x_i,t]\}\frac{\partial}{\partial x_k}E\{C[x_i,t]\}$$

$$=\frac{\partial}{\partial x_k}\left(\phi_{kl}(x_i,t|s)\frac{\partial}{\partial x_l}E\{C[x_i,t]\}\right)+S[x_i,t], \tag{10.60}$$

where the mixed Eulerian –Lagrangian diffusivity is

$$\phi_{kl}(x_i,t|s)=\int_s^t E\{u_k[x_i,t]'u_l(x_i,t|r)''\}dr. \tag{10.61}$$

This "forward" equation for the diffusion of $E\{C[x_i,t]\}$ follows from (10.57), and from the "backward" equation (10.51) for the diffusion of the macro pdf $P(a_l,s|x_j,t)$. That is, the expectation for the present concentration depends upon where the concentration-bearing particles probably were in the past.

If the distinction between $u_l(x_i,t|r)''$ and $u_l(x_i,t|r)'$ is neglected, then $\phi_{kl}=\theta_{kl}$ where

$$\theta_{kl}(x_i,t|s)=\int_s^t E\{u_k[x_i,t]'u_l(x_i,t|r)'\}\,dr. \tag{10.62}$$

With this further approximation, (10.60) becomes the standard "semi-empirical" equation for turbulent diffusion (e.g., Monin and Yaglom, 1971). Similarly, for incompressible flow, the backward diffusion equation (10.51) becomes

$$\frac{\partial}{\partial s}P(a_i,s|x_j,t)+E\{u_k[a_i,s]\}\frac{\partial}{\partial a_k}P(a_i,s|x_j,t)$$

$$=-\frac{\partial}{\partial a_k}\left(\theta_{kl}(a_i,s|t)\frac{\partial}{\partial a_l}P(a_i,s|x_j,t)\right). \tag{10.63}$$

This last semi-empirical equation is jury rigged, as it involves the Lagrangian macro pdf with the Eulerian mean velocity and the mixed Eulerian–Lagrangian diffusivity.

10.5 Reversibility for incompressible flow; the Markov property, Corrsin's hypotheses

The micro distribution for $X_j(a_i,s|t)$ is, again,

$$\pi(a_i,s|x_j,t)=\delta\left(x_j-X_j(a_i,s|t)\right). \tag{10.64}$$

Inverting the label-to-path transformation $(a_i,s)\to(x_j,t)$ yields:

$$\pi(x_j,t|a_i,s)=\delta\left(a_i-X_i(x_j,t|s)\right)=\delta\left(x_j-X_j(a_i,s|t)\right)J_s^t \tag{10.65}$$

where the Jacobi determinant J_s^t of the label-to-path transformation is unity at $t=s$ and is assumed positive at other times. The determinant is identically equal to unity for incompressible flow: $J_s^t=1$ for all t. Hence (Lundgren, 1981) the micro distribution is *reversible*:

$$\pi(a_i, s|x_j, t) = \pi(x_j, t|a_i, s)\,, \tag{10.66}$$

and so is its expectation, the macro distribution P:

$$P(a_i, s|x_j, t) = P(x_j, t|a_i, s)\,. \tag{10.67}$$

Applying (10.67) to (10.63) yields

$$\begin{aligned}&\frac{\partial}{\partial s}P(x_j, t|a_i, s) + E\{u_k[a_i, s]\}\frac{\partial}{\partial a_k}P(x_j, t|a_i, s)\\ &\quad = -\frac{\partial}{\partial a_k}\left(\theta_{kl}(a_i, s|t)\frac{\partial}{\partial a_l}P(x_j, t|a_i, s)\right).\end{aligned} \tag{10.68}$$

The negative sign on the rhs is correct, since the case of interest is $s<t$. If the symbols (a_i, s) and (x_j, t) are now interchanged, (10.68) becomes

$$\begin{aligned}&\frac{\partial}{\partial t}P(a_i, s|x_j, t) + E\{u_k[x_j, t]\}\frac{\partial}{\partial x_k}P(a_i, s|x_j, t)\\ &\quad = \frac{\partial}{\partial x_k}\left(\theta_{kl}(x_j, t|s)\frac{\partial}{\partial x_l}P(a_i, s|x_j, t)\right).\end{aligned} \tag{10.69}$$

Note the change of sign on the rhs as the case of interest is (new) $s<t$, and that there has been a reversal of the integration terminals for θ_{kl}: see (10.62). That is, the macro distribution P, like the tracer concentration C, obeys the semi-empirical equation of turbulent diffusion. Recall the assumption of incompressibility, hence reversibility, and note the absence of a distributed source in (10.69).

Consider statistically stationary turbulence, in which $E\{u_k[x_i, t]\}$ is independent of time, and $\theta_{kl}(x_i, t|s) = \theta_{kl}(x_i|t-s) \sim \theta_{kl}(x_i|\infty)$ as $t-s\to\infty$.[1] Then (10.69) has the asymptotic form

$$\frac{\partial P}{\partial t} \sim \mathcal{D}P \tag{10.70}$$

[1] The flow is not being assumed to be steady; such time translation invariance of moments characterizes statistical stationarity.

where $\mathcal{D}$ is a time-independent linear differential operator with respect to x_k. The asymptotic solution is

$$P \sim e^{(t-s)\mathcal{D}}\delta(x_k - a_k)\,. \tag{10.71}$$

The exponentiated operator should be interpreted as a power series. It readily follows that, asymptotically for large $t-s$, P satisfies the Markov condition (van Kampen, 1992):

$$P(a_i, s|x_j, t) \sim \int P(a_i, s|\xi_k, \tau)P(\xi_l, \tau|x_j, t)dV(\xi) \tag{10.72}$$

for any τ such that $s<\tau<t$. The Lagrangian mean velocity may be expressed as

$$E\{u_k(a_i, s|t)\} = \frac{\partial}{\partial t}E\{X_k(a_i, s|t)\} = \int x_k \frac{\partial}{\partial t}P(a_i, s|x_j, t)dV(x)\,. \tag{10.73}$$

Substituting (10.72) yields

$$E\big\{u_k(a_i, s|t)\big\} \sim \int P(a_i, s|\xi_m, \tau)E\big\{u_k(\xi_l, \tau|t)\big\}dV(\xi)\,, \tag{10.74}$$

which relates Lagrangian mean velocities having different labels. There are analogous expressions relating Lagrangian covariances. These are examples of weak forms of Corrsin's hypotheses (Corrsin, 1959). In the strong form, (10.74) is assumed to hold not asymptotically as $t-\tau \to \infty$, but exactly at $\tau = t$. Then (10.74) becomes

$$E\big\{u_k(a_i, s|t)\big\} = \int P(a_i, s|\xi_m, t)E\big\{u_k[\xi_l, t]\big\}dV(\xi)\,, \tag{10.75}$$

which relates the Lagrangian mean velocity to the Eulerian mean velocity in an intuitively appealing way. The nature of the approximation (10.75) is revealed by a direct approach:

$$\begin{aligned} E\{u_k(a_i, s|t)\} &= E\Big\{u_k\big[X_j(a_i, s|t), t\big]\Big\} \\ &= \int E\Big\{\delta\big(x_l - X_l(a_i, s|t)\big)u_k[x_j, t]\Big\}dV(x) \\ &\sim \int E\Big\{\delta\big(x_l - X_l(a_i, s|t)\big)\Big\}E\big\{u_k[x_j, t]\big\}dV(x) \\ &= \int P(a_i, s|x_l, t)E\big\{u_k[x_j, t]\big\}dV(x)\,. \end{aligned} \tag{10.76}$$

Thus the sole assumption is the statistical independence, at time t, of the Eulerian velocity and the position of a particle released at (a_i, s).

Corrections to (10.75) may be obtained by taking moments of (10.69):

$$E\{u_k(a_i, s|t)\} = \int P(a_i, s|\xi_m, t) E\{u_k[\xi_l, t]\} dV(\xi)$$

$$+ \int \frac{\partial}{\partial \xi_j} \Big(P(a_i, s|\xi_m, t)\theta_{kj}(\xi_l, t|s) \Big) dV(\xi) . \qquad (10.77)$$

Making now the extreme approximation that $P(a_i, s|\xi_l, t) = \delta(a_i - \xi_i)$, which is valid only at $t = s$, yields

$$E\{u_k(a_i, s|t)\} - E\{u_k[a_i, t]\} = \frac{\partial}{\partial a_j}\theta_{kj}(a_i, t|s) . \qquad (10.78)$$

This is an explicit formula for the difference between a Lagrangian mean velocity and an Eulerian mean velocity, that is, for a Stokes drift. Note that the first term on the lhs is the mean of the Lagrangian velocity at time t for particles released at (a_i, s), while the second term is the Eulerian mean velocity at (a_i, t). Expressions similar to (10.78) have been the subject of considerable speculation and numerical experimentation (e.g., Rhines, 1977; Haidvogel and Rhines, 1983).

On the other hand, the simple corollary (1.8) of the labeling theorem (1.4) is an exact result for each realization of a compressible flow.

Finally, rearrangement of the forward diffusion equation (10.69) yields

$$\frac{\partial}{\partial t}P(a_i, s|x_j, t) + \Big(E\{u_k[x_j, t]\} + \frac{\partial}{\partial x_l}\theta^A_{kl}(x_j, t|s) \Big) \frac{\partial}{\partial x_k}P(a_i, s|x_j, t)$$

$$= \frac{\partial}{\partial x_k}\Big(\theta^S_{kl}(x_j, t|s) \frac{\partial}{\partial x_l}P(a_i, s|x_j, t) \Big) \qquad (10.79)$$

where the superscripts A, S denote, respectively, the antisymmetric and symmetric parts of the semi-empirical diffusivity tensor θ_{kl}.

Notes:

(i) the flow has been assumed incompressible;

(ii) for any smooth antisymmetric tensor θ^A_{kl},

$$\frac{\partial^2 \theta^A_{kl}}{\partial x_k \partial x_l} = 0; \qquad (10.80)$$

(iii) the nondivergent vector of divergences of the rows of the antisymmetric part of θ_{kl} contributes to the advective flux of probability, while the symmetric part alone is responsible for the downgradient diffusive flux.

10.6 Scalar concentrations in compressible flow; floats, surface drifters and balloons

An extensive rearrangement of (10.58) yields

$$\frac{\partial}{\partial t} E\{C[x_i, t]\} + \left(E\{u_k[x_i, t]\} + \frac{\partial}{\partial x_l} \phi^A_{kl}(x_i, t|s) + U_k(x_i, t|s) \right)$$
$$\times \frac{\partial}{\partial x_k} E\{C[x_i, t]\} = \frac{\partial}{\partial x_k} \left(\phi^S_{kl}(x_i, t|s) \frac{\partial}{\partial x_l} E\{C[x_i, t]\} \right) + S[x_i, t], \quad (10.81)$$

where

$$U_k(x_i, t|s) = -\int_s^t E\left\{ u_k(x_i, t|r)'' \frac{\partial}{\partial x_l} u_l[x_i, t]' \right\} dr. \quad (10.82)$$

Thus in compressible flow, the Eulerian mean concentration is advected by the Eulerian mean velocity augmented by two drift velocities. The first arises from the inhomogeneity of the skew diffusivity ϕ^A_{kl}, the second from compressibility. A specific form for the sum of these two drifts has been proposed by Gent and McWilliams (1990), and tested in a numerical simulation of the ocean general circulation (Danabosoglu et al., 1994).

Oceanographers deploy subsurface floats in the deep ocean (see Chapter 14). These devices are capable of staying either at a constant depth, or on a surface of constant density. In the first case (isobaric floats), the concentration of floats per unit volume is a scalar in a three-dimensional incompressible flow, thus (10.79) applies. In the second case (isopycnal floats), the float concentration per unit volume, within an approximately isopycnal layer of variable thickness, is a scalar in two-dimensional compressible flow. The layer thickness assumes the role of density, and so (10.81) applies. To explain, the float concentration C satisfies the Lagrangian conservation law

$$\frac{\partial}{\partial t}(hJ^t_s C) = 0, \quad (10.83)$$

while the layer thickness satisfies

$$\frac{\partial}{\partial t}(hJ^t_s) = 0, \quad (10.84)$$

and so

$$\frac{\partial C}{\partial t} = 0. \quad (10.85)$$

Floats which are so buoyant that they remain on the ocean surface are called “drifters” by oceanographers (see Chapter 14). Their concentration C

per unit area of ocean surface satisfies the Lagrangian drifter conservation law in two dimensions:

$$\frac{\partial}{\partial t}(CJ_s^t)=0, \tag{10.86}$$

ignoring acts of piracy. Thus,

$$\frac{\partial L}{\partial t}=-\Delta \tag{10.87}$$

where $L=\ln C$ is the logarithm of concentration and

$$\Delta(a_i, s|t)=\frac{\partial}{\partial t}\ln J_s^t(a_i, s|t)=\frac{\partial}{\partial x_k}u_k\left[X_j(a_i, s|t), t\right] \tag{10.88}$$

is the surface divergence on a particle path.

A backward closure for (10.87) proceeds as in Section 10.4, with the addition here of a source-like term that is correlated to the flow. The procedure leads eventually to (10.81), with $E\{C\}$ replaced by $E\{L\}$, and S replaced by

$$-E\{\Delta[x_i, t]\}+\int_s^t E\left\{u_k[x_i, t]'\frac{\partial}{\partial x_k}\Delta(x_i, t|r)''\right\}dr. \tag{10.89}$$

Thus the Eulerian mean of the log concentration experiences an effective source. Drifters do not spontaneously rise to the ocean surface, and nor do they dive, but they are observed to cluster on the ocean surface around *convergences*.

10.7 Corrections

Diffusion equations were obtained in the preceding sections by making the simplest possible closure approximations to formal expressions for turbulent fluxes. It is instructive to examine higher order corrections. The forward closure of Section 10.3 leads to the linear diffusion equation which is consistent with the CLT in unbounded domains, but which is unsatisfactory in even partially bounded domains. The backward closure of Section 10.4 leads to the standard semi-empirical diffusion equation. This equation is indeed an Eulerian equation, but being standard earns special attention here. Consider for simplicity a passive scalar concentration without a source:

$$\frac{\partial}{\partial t}C(a_j, 0|t)=0, \tag{10.90}$$

subject to the initial condition

$$C(a_j, 0|0)=C_I[a_j, 0]. \tag{10.91}$$

It follows that

$$C[x_j, t] = C_I\left[X_k(x_j, t|0), 0\right], \tag{10.92}$$

and also that

$$\frac{\partial}{\partial t}C[x_j, t] + u_k[x_j, t]\frac{\partial}{\partial x_k}C[x_j, t] = 0, \tag{10.93}$$

subject to

$$C[x_j, 0] = C_I[x_j, 0]. \tag{10.94}$$

The conventional partitioning into means and fluctuations yields

$$\frac{\partial}{\partial t}E\left\{C[x_j, t]\right\} + E\left\{u_k[x_j, t]\right\}\frac{\partial}{\partial x_k}E\left\{C[x_j, t]\right\} = \frac{\partial}{\partial x_k}F_k[x_j, t], \tag{10.95}$$

where the turbulent flux is

$$F_k[x_j, t] = -E\left\{u_k[x_j, t]'C[x_j, t]'\right\}. \tag{10.96}$$

Note the assumption of incompressible flow. The fluctuation $C[x_j, t]'$ obeys

$$\frac{\partial}{\partial t}C[x_j, t]' + u_k[x_j, t]\frac{\partial}{\partial x_k}C[x_j, t]' = \frac{\partial}{\partial x_k}F_k[x_j, t] - u_k[x_j, t]'\frac{\partial}{\partial x_k}E\left\{C[x_j, t]\right\}, \tag{10.97}$$

subject to

$$C[x_j, 0]' = 0. \tag{10.98}$$

The solution of (10.97), (10.98) may be expressed as a time integral from 0 to t, with dummy variable r, along the particle path $X_k(x_j, t|r)$, as in (10.35)–(10.39). In the simple closure approximation leading to the semi-empirical equation (10.60), u_k is replaced with $E\{u_k\}$ in the argument of F_k along the path. Consequently, this part of the Lagrangian solution for C' makes no contribution through (10.96) to the mean flux F_k. Also, replacing r with t in the argument of $E\{C\}$ causes $X_k(x_j, t|r)$ to be replaced with x_k, yielding the semi-empirical downgradient formula for concentration diffusion. These crude approximations can be refined. First, the flux divergence in the time integral solution of (10.97), (10.98) may be expanded about $E\{X_k\}$ in powers of $X_k{}'$, while $E\{C[X_k, t]\}$ may be replaced with $E\left\{C\left[E\{X_k\}, t\right]\right\}$ instead of $E\{C[x_k, t]\}$. These better approximations lead to the following implicit equation for the flux:

$$F_k[x_j, t] - \int_0^t \phi_{km}^S(x_j, t|r) \frac{\partial^2 F_n}{\partial x_m \partial x_n} \left[E\{X_l(x_j, t|r)\}, r \right] dr$$
$$= -\int_0^t E\{u_k[x_j, t]' u_m(x_j, t|r)''\} \frac{\partial E\{C\}}{\partial x_m} \left[E\{X_l(x_j, t|r)\}, r \right] dr. \tag{10.99}$$

Davis (1987) obtained a closure equating F_k directly to an integral somewhat similar to the rhs of (10.99). Consider what has been neglected in deriving (10.99). One approximation is

$$F_k[X_j, r] = F_k[E\{X_k\}, r] + X_m{}' \frac{\partial F_k}{\partial x_m}[E\{X_k\}, r] + \mathcal{O}\left(\left(X_m{}' \frac{\partial}{\partial x_m} \right)^2 F_k \right). \tag{10.100}$$

If $X_k{}' \sim \lambda$ and $\partial F_k / \partial x_m \sim \mathcal{F}/\mathcal{L}$, then the relative error in (10.100) is $\lambda/\mathcal{L}$. The other approximation is, in effect,

$$E\{u_k(x_j, t|r)' u_m(x_j, t|r)'' u_n(x_j, t|w)'\} \ll E\{(u')^2\}^{3/2} \tag{10.101}$$

which can hold even if the Lagrangian velocity covariance does not vanish at large lag. That is, (10.101) is an assumption about a third moment, and hence about the symmetry of the pdf for velocity.

Now consider the simplification of (10.99) to the semi-empirical gradient formula

$$F_k[x_j, t] = -\phi_{kl}^S(x_i, t|0) \frac{\partial}{\partial x_l} E\{C[x_m, t]\}. \tag{10.102}$$

If the scale of the diffusivity is $\phi_{kl}^S \sim \Phi$, then the diffusion time for mean concentration fields is $\mathcal{T} \sim \mathcal{L}^2 \Phi^{-1}$, so the second term on the lhs of (10.99) is not negligible. If the mixed Eulerian–Lagrangian velocity covariance on the rhs of (10.99) has the decorrelation time scale τ, then $|x_i - E\{X_i\}| \sim \tau U$ where $|E\{u_j\}| \sim U$, and the relative error in the rhs of (10.102) is $O(\tau U \mathcal{L}^{-1}) = O((\lambda \mathcal{L}^{-1}) U \mathcal{U}^{-1})$, since $\lambda \sim \mathcal{U}\tau$ where $E\{(u_j{}')^2\} \sim \mathcal{U}^2$. In the ocean, $\lambda \sim$ 50 km, $\mathcal{L} \sim 500$ km, $U \sim 10^{-2}\,\mathrm{m\,s^{-1}}$, while $\mathcal{U} \sim 10^{-1}\,\mathrm{m\,s^{-1}}$. Hence, lagging the mean concentration gradient as in (10.99) yields only a 1% correction. However, $E\{X_l(x_j, t|r)\}$ on the lhs of (10.99) cannot be replaced with x_l, since $\phi_{km}^S(x_j, t|r) \not\to 0$ as $t - r \to \infty$. In summary, the lhs of (10.99) should be equated to the rhs of (10.102). The resulting equation for $E\{C\}$ is a higher order diffusion equation with mean Lagrangian arguments. However, it may make more sense to use only the simplest closure. If the turbulence is homogeneous, and if the single-point Lagrangian velocity field is jointly Gaussian at different times, then the *infinite* series of corrections may be summed. The result is the simple gradient formula for the flux. To see this, reconsider the Lagrangian solution for the concentration:

$$C[x_i, t] = C_I\left[x_i - \int_0^t u_i(x_j, t|r)dr, 0\right]. \qquad (10.103)$$

Assuming, without loss of generality for homogeneous turbulence, that $E\{u_i(x_j, t|r)\} = 0$, it follows that

$$E\{C[x_i, t]\} = \exp\left(\frac{1}{2}E\{(X_k - x_k)(X_j - x_j)\}\frac{\partial^2}{\partial x_k \partial x_j}\right) C_I[x_i, 0], \qquad (10.104)$$

where $X_k = X_k(x_j, t|0)$ and the covariance is homogeneous, that is, independent of x_i. Now (10.104) is the exact solution of the simple diffusion equation

$$\frac{\partial}{\partial t}E\{C[x_i, t]\} = \theta^S_{kj}(t|0)\frac{\partial^2}{\partial x_k \partial x_j}E\{C[x_i, t]\} \qquad (10.105)$$

subject to the initial condition (10.94). In conclusion, it would seem risky to add a finite number of corrections to the simplest closure of inhomogeneous nonstationary non-Gaussian turbulence.

10.8 Random flight models and plankton dynamics

The linear diffusion equation (10.30), for the pdf of $X_k(a_j, s|t)$, holds asymptotically for large $t - s$. It holds exactly, for all $t > s$, for the pdf of the solution of the stochastic differential equation or SDE

$$dX_k(a_j, s|t) = E\{u_k(a_j, s|t)\}dt + 2^{1/2}\left(\kappa^{1/2}(a_j, s|t)\right)_{kl} d\beta_l(t), \qquad (10.106)$$

where $(\kappa^{1/2})_{nm}$ is an element of the square root of the symmetric part of the matrix of Taylor diffusivities κ_{nm}:

$$(\kappa^{1/2})_{nm}(\kappa^{1/2})_{ml} = \kappa^S_{nl}, \qquad (10.107)$$

and the increments $d\beta_k$ are mutually independent Wiener processes, that is,

$$E\{d\beta_k(t)\} = 0, \quad E\{d\beta_j(t)d\beta_k(t')\} = \delta_{jk}\delta(t - t'). \qquad (10.108)$$

See, e.g., Gardner (1985) and Rodean (1996). The square root matrix is real if the Taylor diffusivity matrix has only nonnegative eigenvalues. The initial condition for (10.106) is simply $X_k(a_j, s|s) = a_k$. The SDE is linear since the coefficients are independent of X_k, and so the solution for X_k in an unbounded domain has a Gaussian distribution.

The semi-empirical equation (10.69) is exactly the Fokker–Planck equation (FPE) for the Itō SDE

$$dx_j(t) = \Big(E\{u_j[x_k(t), t]\} + \frac{\partial}{\partial x_l}\theta_{jl}(x_k(t), t|s)\Big)dt + 2^{1/2}\big(\theta^{1/2}(x_k(t), t|s)\big)_{jl} d\beta_l(t), \tag{10.109}$$

where $(\theta^{1/2})_{nm}$ is the matrix square root of the symmetric part of the matrix θ_{nm} (Gardner, 1985; Rodean, 1996). Recall that (10.69) assumes incompressible flow. The Itō SDE (10.109) is nonlinear, since the coefficients depend upon the solution $x_k(t)$. Solutions are not in general Gaussian.

If the divergence of the diffusivity is omitted from the "drift" coefficient in the Itō SDE (10.109), that is, if

$$dx_j(t) = E\{u_j[x_k(t), t]\}dt + 2^{1/2}\big(\theta^{1/2}(x_k(t), t|s)\big)_{jl} d\beta_l(t), \tag{10.110}$$

then the corresponding FPE for the pdf of $x_j(t)$ is

$$\frac{\partial}{\partial t}P(a_i, s|x_j, t) + \frac{\partial}{\partial x_k}\Big(E\{u_k[x_j, t]\}P(a_i, s|x_j, t)\Big) = \frac{\partial^2}{\partial x_k \partial x_l}\Big(\theta_{kl}(x_j, t|s)P(a_i, s|x_j, t)\Big). \tag{10.111}$$

Yamazaki and Kamykowski (1991) use (10.110), in one dimension, to model vertical trajectories of motile phytoplankton in a wind-mixed water column. Holloway (1994) points out, in effect, that the corresponding FPE (10.111) differs from the semi-empirical equation (10.69) for turbulent diffusion. So long as the diffusivity is nonuniform, such "wind mixing" would cause an initially uniform distribution to become unrealistically nonuniform. Yamazaki and Kamykowski (1994) reply that the derivation of (10.69) assumes incompressibility of the flow, which could only be satisfied by a one-dimensional flow if it were unrealistically uniform in the vertical. Perhaps the resolution is that one must assume incompressible, three-dimensional flow (with $x_3 = z$, $a_3 = c$, $u_3 = w$, $\theta_{33} = \theta$, $\beta_3 = \beta$, etc.), but seek a FPE for the marginal pdf P for the vertical: $P = P(c, s|z, t)$. This is possible if both E{w} and θ depend only upon the vertical: $E\{w\} = E\{w[z, t]\}$, $\theta = \theta(z, t|s)$. Then (10.69) becomes

$$\frac{\partial}{\partial t}P + \frac{\partial}{\partial z}(E\{w\}P) = \frac{\partial}{\partial z}\Big(\theta\frac{\partial}{\partial z}P\Big) \tag{10.112}$$

which is the FPE for the Itō SDE

$$dz(t) = \big(E\{w[z(t), t]\} + \frac{\partial}{\partial z}\theta(z(t), t|s)\big)dt + \sqrt{2\theta(z(t), t|s)}\,d\beta(t). \tag{10.113}$$

The real issue is the validity of the semi-empirical equation of turbulent diffusion (10.69), even for incompressible flow.

The stochastic processes (10.106), (10.109) and (10.113) for the particle path $X_k(a_j, s|t)$ are Markovian, yet this need not be true of the Lagrangian displacement (10.2) which they model. For example, the exact result (10.105) is consistent with a Markovian model, even though the Gaussian homogeneous and stationary velocity field in (10.103) may be otherwise: an *anticipating* process, for example (Gardner, 1985, p. 86; Rodean, 1996, p. 28). Analogous to the stochastic processes for the particle path $X_k(a_j, s|t)$ are Langevin equations for the particle velocity $u_k(a_j, s|t)$. The coefficient of dt is the expectation of the Lagrangian acceleration. The joint pdfs for Lagrangian velocities and passive scalars may be estimated from numerical solutions of these equations (Pope, 1994). Again the implicit assumption is that the Lagrangian fields are Markov processes in time.

10.9 Annual plankton patchiness

Consider plankton concentration C, subject to advection by incompressible planar flow u_k, and growing along the particle path $x_j = X_j(a_k, s|t)$ according to a Lagrangian logistic model

$$\frac{\partial}{\partial t}C = rC\left(1 - \frac{C}{C_p}\right), \tag{10.114}$$

where $C = C(a_j, s|t)$. The growth rate is $r = r(a_j, s|t) = r[X_i(a_j, s|t), t]$ where the Eulerian field $r[x_i, t]$ is specified, and similarly for the carrying capacity C_p. It is convenient (Levins, 1969) to introduce the nonlinearly transformed variable

$$D = \ln\left(C(C_p - C)^{-1}\right), \tag{10.115}$$

which ranges from $-\infty$ to ∞. Then (10.114) becomes a simple conservation law with a source:

$$\frac{\partial}{\partial t}D = r. \tag{10.116}$$

The unbounded linear growth permitted by (10.116), in the case of constant r, represents an approach by the original concentration C to either of its finite limits 0 or C_p.

Assuming that the velocity field u_j is incompressible, and turbulence that is stationary, homogeneous and isotropic, and assuming without loss of

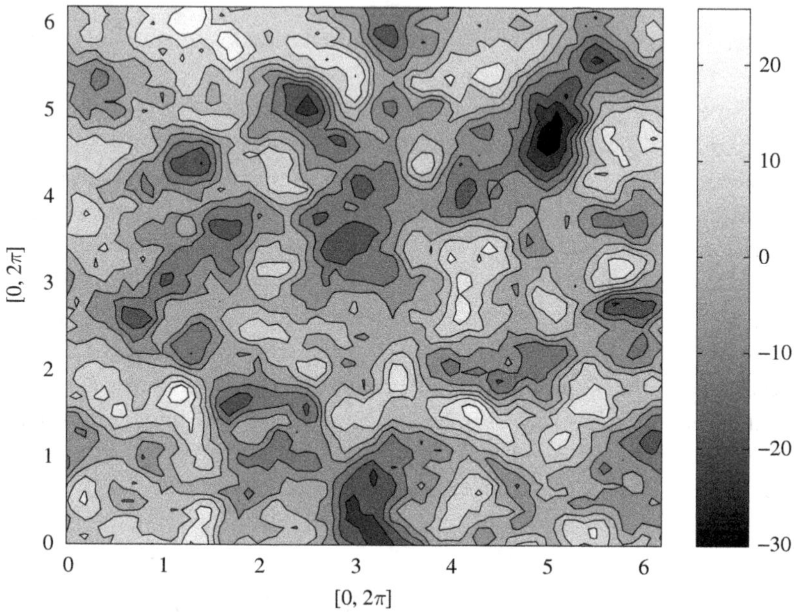

Figure 10.1 Contours of the amplitude of $E\{D\}$, the expectation of nonlinearly transformed plankton concentration, at the annual period T. The amplitude is derived from (10.117), after separating variables in space and time, for one realization of the amplitude of a spatially uncorrelated plankton growth rate r at the annual period. The expectation is taken with respect to the ensemble of random flows; the amplitudes are real and may be positive or negative. A dimensionless doubly periodic domain $[0, 2\pi] \times [0, 2\pi]$ is chosen for computational convenience alone. The concentration patches have the length-scale $L = \sqrt{2\pi\phi T}$, which is independent of domain width. For the scale values given in the text, $L = 700\,\mathrm{km}$. In this computation, the width of the patches is about one quarter of the width of the domain, thus the latter is about 2800 km. It is emphasized that patches of spatial scale L would emerge regardless of the width of the domain; L is the distance that mean transformed concentration can diffuse in one year. After Bennett and Denman (1989)

generality that the expectation velocity vanishes, the semi-empirical diffusion equation (10.60) for the expectation of the transformed concentration is[2]

$$\frac{\partial}{\partial t} E\{D[x_k, t]\} = \phi \frac{\partial^2}{\partial x_j \partial x_j} E\{D[x_k, t]\} + r[x_k, t], \tag{10.117}$$

[2] The advantage of closing the turbulent fluxes after transforming the concentration was pointed out to the author by G. Holloway (personal communication, 1985).

where ϕ is the turbulent diffusivity. It is immediately obvious that even if the growth rate r is white noise in space, an annual cycle in its amplitude will lead to patchiness in the expectation of the transformed concentration D, with a length-scale $L = \sqrt{2\pi\phi T}$ where T is one year. For example, an rms velocity of 0.05 m s^{-1} and a decorrelation length of 5×10^4 m implies $\phi = 0.25 \times 10^4\,\mathrm{m^2 s^{-1}}$ and hence $L = 7 \times 10^5$ m. This is the distance which the transformed concentration can diffuse in one year (Bennett and Denman, 1989). An example of dispersion-induced annual patchiness is shown in Figure 10.1. The patch scale $L = 700$ km is independent of the domain width, which is about 2800 km in this doubly periodic simulation.

11
Relative dispersion

11.1 Joint displacement of a pair of particles

Consider a pair of fluid particles occupying the two points a_i, b_j at time s. Their subsequent positions at time t are $x_n = X_n(a_i, s|t)$, $y_m = X_m(b_j, s|t)$; see Figure 11.1.

> *Aside*: It is now convenient to distinguish between the point x_n and the functional form X_n; otherwise the position of the second particle would have to be denoted $Y_m(b_j, s|t)$ even though the path function is the same for both particles.

The "micro joint pdf" for the positions of the two particles is

$$\pi(a_i, b_j, s|x_n, y_m, t) \equiv \delta\big(x_n - X_n(a_i, s|t)\big)\delta\big(y_m - X_m(b_j, s|t)\big) \tag{11.1}$$

where for example

$$X_n(a_i, s|t) = a_n + \int_s^t u_n(a_i, s|w)dw\,. \tag{11.2}$$

> *Aside*: In this section and in other discussions of particle separation, the symbols r_n and r will be reserved for the spatial separation vector and its magnitude, respectively. In particular r will not denote a running time variable between s and t. The symbols v, w will be reserved for that purpose.

The labeling theorem for two particles yields

$$\left(\frac{\partial}{\partial s} + u_k[a_i, s]\frac{\partial}{\partial a_k} + u_l[b_j, s]\frac{\partial}{\partial b_l}\right)\pi(a_i, b_j, s|x_n, y_m, t) = 0\,. \tag{11.3}$$

A backward closure approximation may be made just as in Section 10.4, yielding a diffusion-like equation for the macro joint pdf P, where the all-purpose symbol P now denotes

$$P(a_i, b_j, s|x_n, y_m, t) = E\big\{\pi(a_i, b_j, s|x_n, y_m, t)\big\}\,. \tag{11.4}$$

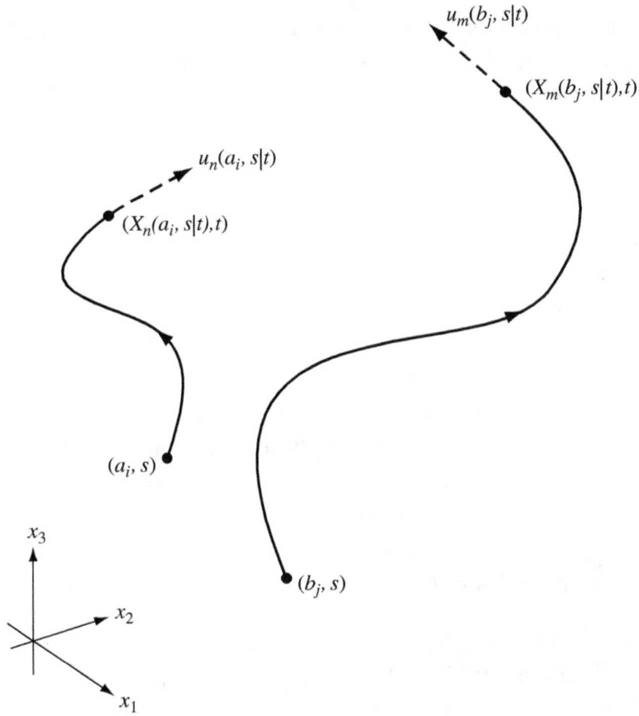

Figure 11.1 A pair of particles labeled by their release positions a_i and b_j at time s.

The assumption of incompressibility leads as before to a structural simplification of the diffusion-like equation, and also permits reversal of the *arguments* of both π and P. For example,

$$P(x_n, y_m, t|a_i, b_j, s) = P(a_i, b_j, s|x_n, y_m, t), \tag{11.5}$$

where a_i, b_j, s are the independent variables in (11.3).

Reversing *notation* (so that x_n, y_m, t are the independent variables in (11.3)), and taking the case of interest to be (new) $s < t$, the simple closure becomes

$$\begin{aligned}
&\left(\frac{\partial}{\partial t} + E\{u_k[x_n, t]\}\frac{\partial}{\partial x_k} + E\{u_l[y_m, t]\}\frac{\partial}{\partial y_l}\right) P(a_i, b_j, s|x_n, y_m, t) \\
&= \frac{\partial}{\partial x_k}\left(\phi_{kp}(x_n, x_n, t|s)\frac{\partial}{\partial x_p} + \phi_{kq}(x_n, y_m, t|s)\frac{\partial}{\partial y_q}\right) P(a_i, b_j, s|x_n, y_m, t) \\
&\quad + \frac{\partial}{\partial y_l}\left(\phi_{lp}(y_m, x_n, t|s)\frac{\partial}{\partial x_p} + \phi_{lq}(y_m, y_m, t|s)\frac{\partial}{\partial y_q}\right) P(a_i, b_j, s|x_n, y_m, t).
\end{aligned} \tag{11.6}$$

Note that the mean advecting velocities on the left-hand side (lhs) of (11.6) are solenoidal (incompressible flow). The diffusivities on the right-hand side (rhs) of (11.6) are

$$\phi_{kp}(x_n, y_m, t|s) = \int_s^t E\{u_k[x_n, t]' u_p(y_m, t|w)''\} dw. \tag{11.7}$$

In particular, the single-particle diffusivity $\phi_{kp}(x_n, t|s)$ of Section 10.4 is the same as $\phi_{kp}(x_n, x_n, t|s)$. Note that no summation is implied by repeated indices in argument lists. Recall that $u_p(x_n, t|s)'' = u_p[\xi_k, s]'$ evaluated at $\xi_k = X_k(x_n, t|s)$. Lundgren (1981) deduces equation (11.6) in a superficially different manner. However, (11.6) is unacceptable because the marginal equation, obtained by integrating (11.6) with respect to the position of the second particle, is not the same as the equation (10.69) for the marginal single-particle pdf $P(a_i, s|x_n, t)$. For example, if (11.6) were used to compute the drift of one particle, the result would depend upon the presence of the second particle. It is also readily shown that (11.6) predicts a nonzero mean *vector* separation rate for pairs of particles in homogeneous turbulence, which is absurd.

Exercise 11.1 Integrate (11.6) with respect to y_m, over an unbounded domain. Compare the result with (10.69). □

Exercise 11.2 Use (11.6) to determine $\partial E\{x_n\}/\partial t$, where the mean is taken over all x_n and y_m in an unbounded domain. □

Exercise 11.3 Use (11.6) to determine $\partial E\{x_n - y_n\}/\partial t$ in an unbounded domain. □

The shortcomings of (11.6) may be remedied. Noting that the difficulties arise from the position of the inner partial derivative in the mixed terms on the rhs, consider that

$$\begin{aligned}\frac{\partial}{\partial y_q} E\{u_k[x_n, t]' u_q(y_m, t|w)''\} &= E\left\{u_k[x_n, t]' \left(\frac{\partial u_q'}{\partial \xi_p}\right)(y_m, t|w) \frac{\partial}{\partial y_q} X_p(y_m, t|s)\right\} \\ &\cong E\left\{u_k[x_n, t]' \left(\frac{\partial u_q'}{\partial \xi_p}\right)(y_m, t|w)\right\} \\ &\quad \times E\left\{\frac{\partial}{\partial y_q} X_p(y_m, t|s)\right\},\end{aligned} \tag{11.8}$$

provided we neglect triple correlations, which is consistent with the approximate closure leading to (11.6). Moreover, if the turbulence is locally homogeneous, then the rhs of (11.8) is approximately

$$E\left\{u_k[x_n,t]'\left(\frac{\partial u_q'}{\partial \xi_p}\right)(y_m,t|w)\right\}\delta_{pq}=E\left\{u_k[x_n,t]'\left(\frac{\partial u_p'}{\partial \xi_p}\right)(y_m,t|w)\right\}=0 \tag{11.9}$$

for incompressible flow. Consequently, the mixed terms in (11.6) may be replaced with more satisfactory forms, for example

$$\begin{aligned}&\phi_{kq}(x_n,y_m,t|s)\frac{\partial}{\partial y_q}P(a_i,b_j,s|x_n,y_m,t)\\&\Rightarrow\frac{\partial}{\partial y_q}\left(\phi_{kq}(x_n,y_m,t|s)P(a_i,b_j,s|x_n,y_m,t)\right).\end{aligned} \tag{11.10}$$

Then the marginal equation for a single particle is the same as (10.69). In fact, for weakly inhomogeneous turbulence, *all* the derivatives on the rhs of (11.6) may be moved to the left of the diffusivities. The resulting equation is the same as that of Kraichnan (1965), except that in the latter only the solenoidal parts of the Lagrangian velocities appear in the diffusivities. For example, $u_k(x_n,t|w)''$ is arbitrarily replaced with $u_k^s(x_n,t|w)''$, where by definition

$$\frac{\partial}{\partial x_k}u_k^s(x_n,t|w)''=0. \tag{11.11}$$

Of course, such a replacement also permits moving the derivatives in the desired manner.

The approximate equation (11.6) is presumably invalid for strongly inhomogeneous turbulence. Such a flow may be characterized by a mean strain rate Λ greatly in excess of the root mean square strain rate. Hence the displacement of a particle by the mean flow, through the eddy field, grows as $\exp\left[\Lambda(t-s)\right]$. Pairs of particles initially within the same eddy will be moving independently as soon as $t-s\sim|\Lambda|^{-1}$, and so

$$P(a_i,b_j,s|x_n,y_m,t)\sim P(b_j,s|y_n,t)P(b_j,s|y_m,t), \tag{11.12}$$

which obviates the need for an evolution equation for the two-particle pdf.

Exercise 11.4 Explain the middle term in the triple products in (11.8). □

11.2 Separation of a pair of particles

Specializing to the case of incompressible isotropic turbulence, it is clear that the diffusivities in (11.6) are solenoidal with respect to either index and so the rearrangements exemplified by (11.10) are exactly valid.

Exercise 11.5 Assuming that

$$\frac{\partial u_i}{\partial x_i}[x_j, t]=0,$$

and that

$$\phi_{kq}(x_n, y_m, t|s)=\int_s^t E\{u_k[x_n, t]'u_q(y_m, t|w)''\}dw=\phi_{kq}(x_n-y_m, t|s), \tag{11.13}$$

verify that

$$\frac{\partial}{\partial x_k}\phi_{kq}=\frac{\partial}{\partial y_q}\phi_{kq}=0. \tag{11.14}$$

□

Let centroid and separation coordinates for particle pairs be defined by

$$c_j=\frac{1}{2}(a_j+b_j), \quad f_j=b_j-a_j, \text{ at time } s \tag{11.15}$$

$$z_n=\frac{1}{2}(x_n+y_n), \quad r_n=y_n-x_n, \text{ at time } t; \tag{11.16}$$

see Figure 11.2.

Let the pair pdf in centroid and separation coordinates be $P(c_i, f_j, s|z_m, r_n, t)$. The marginal pdf for vector separation:

$$P(f_j, s|r_n, t)=\int P(c_i, f_j, s|z_m, r_n, t)dV(z), \tag{11.17}$$

is independent of the initial centroid c_i by homogeneity. It is straightforward to derive from (11.6), (11.10) etc., the following evolution equation for the marginal pdf for vector separation:

$$\frac{\partial}{\partial t}P(f_j, s|r_n, t)=\frac{\partial^2}{\partial r_k \partial r_l}\left(\eta_{kl}(r_n, t|s)P(f_j, s|r_n, t)\right), \tag{11.18}$$

where the effective diffusivity is

$$\eta_{kl}(r_n, t|s)=2\left(\phi_{kl}(x_n, x_n, t|s)-\phi_{kl}(x_n, x_n+r_n, t|s)\right), \tag{11.19}$$

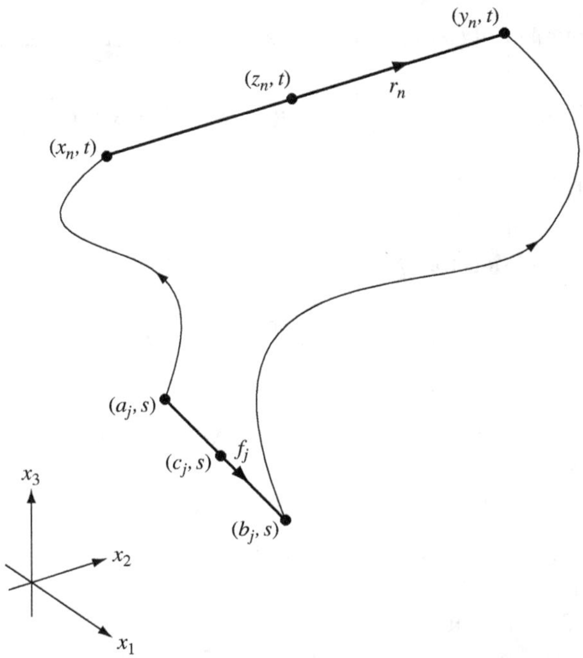

Figure 11.2 Centroid coordinates c_j, z_n and separation coordinates f_j, r_n for the particle pair.

again assuming homogeneity. The initial condition for (11.18) is

$$P(f_j, s|r_n, s) = \delta(r_n - f_n). \tag{11.20}$$

Next, average the vector separation pdf over the direction of f_j, that is, over the direction of separation at time s. The result depends only upon the magnitude $f = \sqrt{f_j f_j}$ of the separation at time s and, as a consequence of isotropy, only upon the magnitude $r = \sqrt{r_n r_n}$ of the separation at time t:

$$P(f, s|r, t) = A(f)^{-1} \int P(f_j, s|r_n, t) dA(f), \tag{11.21}$$

where $dA(f)$ is an elemental area on a surface of radius f, and $A(f)$ is the area of that surface:

$$\begin{aligned} A(f) &= 2\pi f, \quad D = 2 \\ A(f) &= 4\pi f^2, \quad D = 3, \end{aligned} \tag{11.22}$$

D being the number of spatial dimensions. The pdf for scalar separation obeys

$$\frac{\partial}{\partial t}P(f,s|r,t)=A(r)^{-1}\frac{\partial}{\partial r}\left\{A(r)\eta(r,t|s)\frac{\partial}{\partial r}P(f,s|r,t)\right\}, \tag{11.23}$$

where $\eta(r,t|s)$ is the longitudinal component of the incompressible isotropic diffusivity tensor $\eta_{jk}(r,t|s)$.

The initial condition for (11.23) is

$$P(f,s|r,s)=A(f)^{-1}\delta(r-f). \tag{11.24}$$

Exercise 11.6 Derive (11.23) from (11.18)–(11.20), paraphrasing Batchelor (1960, Chapter 3). □

Richardson (1926) virtually guesses (11.23), although he does not consider time dependence for η. Kraichnan (1965, 1966a, 1966b, equation [3.6]) derives (11.23) using his "Abridged Lagrangian History Direct Interaction Approximation." Lundgren (1981) derives (11.23) by only assuming a velocity field delta correlated in time. That assumption is equivalent to the approximations (i) and (ii) made here in Section 10.4. By the arguments of Batchelor (1960), the longitudinal diffusivity η is expressible as

$$\eta(r,t|s)=r^{-D}\int_0^r \zeta(\rho,t|s)\rho^{D-1}d\rho, \tag{11.25}$$

where ζ is the relative diffusivity defined by

$$\zeta(r,t|s)=E\{\frac{\partial}{\partial t}r^2\}\Big|_r=\eta_{kk}(r,t|s), \tag{11.26}$$

the expectation being conditioned by the separation having the value r at time t. Finally, note that

$$\int_0^\infty A(r)P(f,s|r,t)dr\equiv 1, \tag{11.27}$$

provided

$$A(r)\eta(r,t|s)\frac{\partial}{\partial r}P(f,s|r,t)\to 0 \tag{11.28}$$

as $r\to 0$, and as $r\to\infty$.

11.3 Richardson's self-similar asymptotic solution

Observations in the lower atmosphere led Richardson (1926) to infer that the longitudinal diffusivity is steady, and depends algebraically on the separation:

$$\eta = \beta r^{4/3}, \tag{11.29}$$

where β is a (positive) constant having dimensions of $(\text{length})^{2/3} \times (\text{time})^{-1}$; see Figure 11.3.

Exercise 11.7 Consider three-dimensional flow ($D=3$). Adopting Richardson's law (11.29), show that the initial value problem (11.23), (11.24) for the separation pdf has the solution

$$P(f, s|r, t) = 3\left(8\pi\theta f^{7/6} r^{7/6}\right)^{-1} \exp\left(-9\left(f^{2/3}+r^{2/3}\right)(4\theta)^{-1}\right) \times I_{7/2}\left(9 f^{1/3} r^{1/3} (2\theta)^{-1}\right), \tag{11.30}$$

where $\theta = \beta(t-s) > 0$ and $I_{7/2}$ is a modified Bessel function of the first kind; see Figure 11.4.

Hint: take the Laplace transform in time of (11.23), subject to (11.24), where η is given by (11.29). Verify that the complementary function associated with the resulting ordinary differential equation for $\overline{P}$, the transform of P, is

$$\overline{P} = A r^{-7/6} I_{7/2}\left(3q^{1/2} r^{1/3}\right) + B r^{-7/6} K_{7/2}\left(3q^{1/2} r^{1/3}\right), \tag{11.31}$$

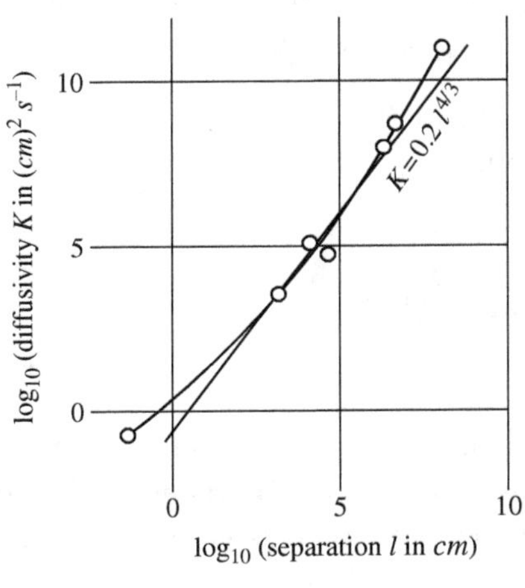

Figure 11.3 Common logarithm of relative diffusivity K (here, $E\{\zeta\}$) in $(cm)^2 s^{-1}$ *versus* common logarithm of separation l (here, r) in cm; after Richardson (1926).

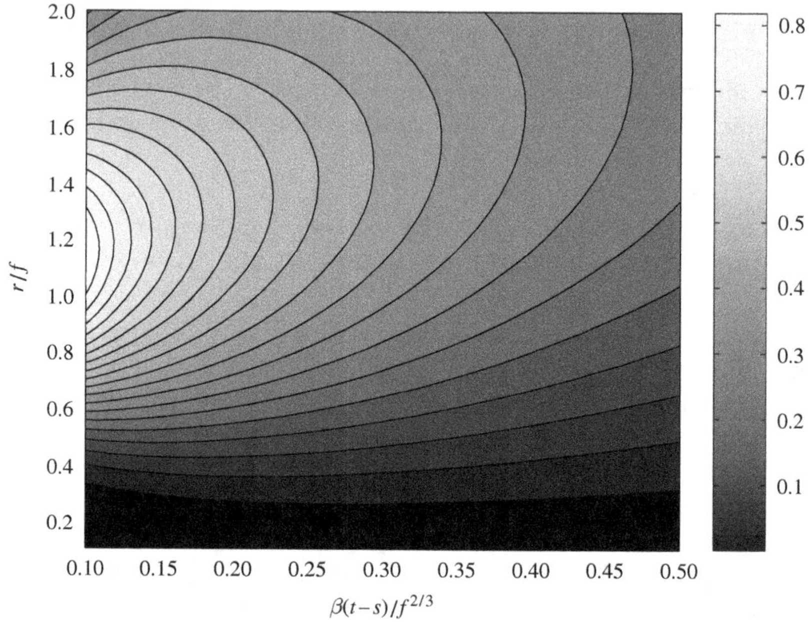

Figure 11.4 Measure-weighted, three-dimensional separation pdf $4\pi r^2 fP$ after Richardson (1926), as a function of scaled time $\beta(t-s)/f^{2/3}$ and scaled separation r/f; see (11.30).

where $K_{7/2}$ is a modified Bessel function of the third kind, A and B are constants, and q is the Laplace transform variable.

Next, show that in the limit as $r^{2/3}/\theta \to 0$ but $r \gg f$, (11.30) has the self-similar asymptotic form

$$P(f,s|r,t) \sim \left(4\pi\Gamma\left(\frac{9}{2}\right)\theta^{9/2}\right)^{-1}\left(\frac{3}{2}\right)^{4}\exp\left(-9r^{2/3}(4\theta)^{-1}\right), \qquad (11.32)$$

where Γ is the gamma function; see Figure 11.5. This form, which is independent of the initial separation f, is the 3D variant of Richardson's (1926) solution in 2D. Note the normalization implied in (11.24). Finally, show that for any n,

$$E\{r^n\} \sim c\theta^{3n/2}, \qquad (11.33)$$

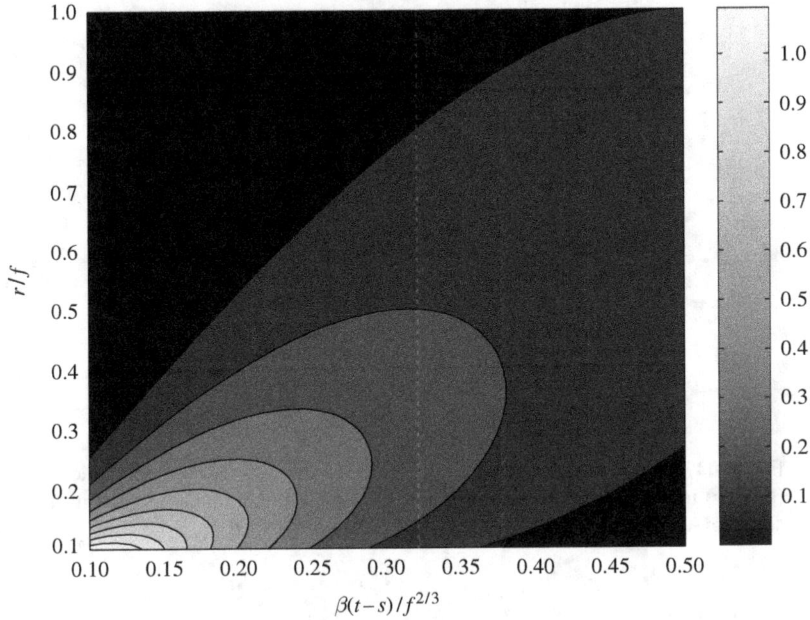

Figure 11.5 Measure-weighted, three-dimensional separation pdf $4\pi r^2 fP$ after Richardson (1926), here the self-similar asymptotic form in the limit $\beta(t-s)/f^{2/3} \to \infty$, $r/f \to \infty$; see (11.32).

where c is a dimensionless constant. In particular, the kurtosis

$$\mathrm{kurt}(r) = \frac{E\left\{(r - E\{r\})^4\right\}}{\left(E\left\{(r - E\{r\})^2\right\}\right)^2} \tag{11.34}$$

is a dimensionless constant, asymptotically for large time. □

11.4 Lundgren's log normal solution

There is evidence in the upper atmosphere (Morel and Larcheveque, 1974; see Figure 11.6) and in the oceanic main thermocline (Price, cited in McWilliams et al., 1983; see Figure 11.7) for approximately two-dimensional turbulence having longitudinal diffusivities of the form

$$\eta = T^{-1} r^2, \tag{11.35}$$

where T is a constant with the dimension of time.

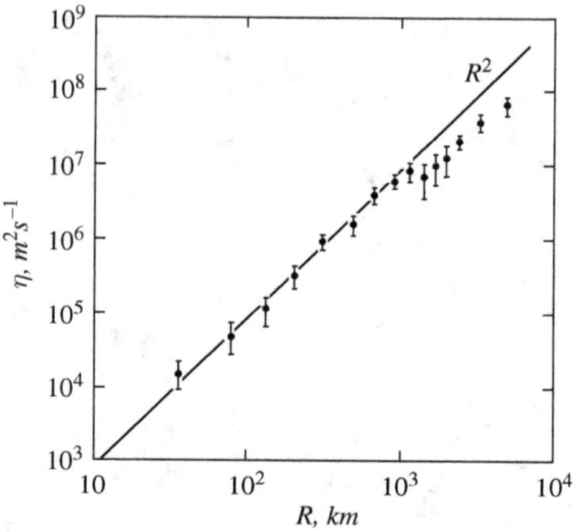

Figure 11.6 Relative diffusivity η (here, $E\{\zeta\}$) in $\mathrm{m}^2\ \mathrm{s}^{-1}$ *versus* separation R (here, r) in km, inferred from observations of high-altitude balloon pairs in the southern hemisphere; after Morel and Larcheveque (1974).

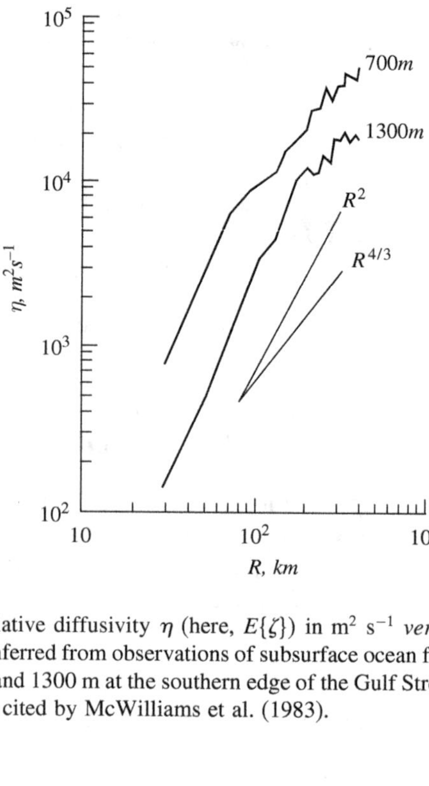

Figure 11.7 Relative diffusivity η (here, $E\{\zeta\}$) in $\mathrm{m}^2\ \mathrm{s}^{-1}$ *versus* separation R (here, r) in km, inferred from observations of subsurface ocean floats at drifters at depths of 700 m and 1300 m at the southern edge of the Gulf Stream recirculation gyre; after Price, cited by McWilliams et al. (1983).

The r^2 dependence is clear for atmospheric (oceanic) separations r in the range 30–1000 km (30–100 km); for separations above 1000 km (100 km) it is less well defined.[1] The corresponding solution ($D=2$) of (11.23), (11.24) is log normal (Lundgren, 1981):

$$P(f, s|r, t) = (4\pi r^2 \sigma)^{-1/2} \exp\left(-\frac{(L+2\sigma)^2}{4\sigma}\right), \tag{11.36}$$

where $L = \ln(r/f)$ and $\sigma = T^{-1}(t-s)$; see Figure 11.8.

Exercise 11.8 Show that

$$E\{r^n\} = f^n e^{n(n+2)\sigma}, \tag{11.37}$$

for $-\infty < n < \infty$. There is exponential growth for $n < -2$ and for $n > 0$, and decay otherwise! In particular, the kurtosis grows as $\exp(8\sigma)$, asymptotically for large σ. □

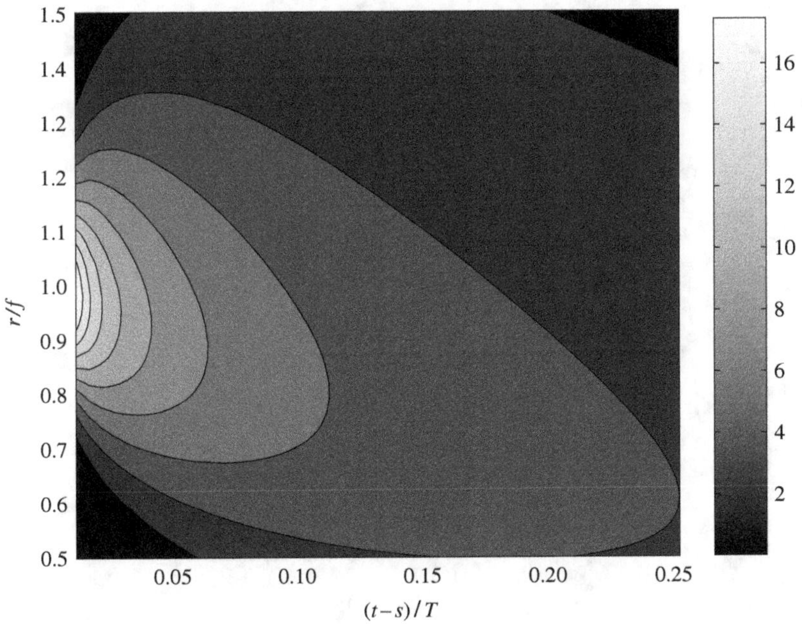

Figure 11.8 Measure-weighted, two-dimensional separation pdf $2\pi r P$ after Lundgren (1981), as a function of scaled time $(t-s)/T$ and separation r/f; see (11.36).

[1] LaCasce and Bower (2000) review several sets of North Atlantic subsurface float data, but find no simple dependence of η upon r; their statistics tend to favor a 4/3 power law.

11.5 Observations of dispersion

There have been few attempts to observe the separation pdf $P(f, s|r, t)$. Dye measurements (Figure 11.9) in Lake Huron (Sullivan, 1971) do not support Richardson's solution of (11.23), (11.24) based on the assumption that $\eta \propto r^{4/3}$. The measurements are more consistent with a normal distribution for the vector separation r_j, which may be derived from (11.18) by assuming that η_{jk} is independent of r_j. This would be the case if the two particles were moving independently with normally distributed displacements, that is, for an elapsed time greatly exceeding the decorrelation time of the turbulent velocity field.

The pdfs for the zonal and meridional components of separation of high-altitude balloons are estimated by Er-El and Peskin (1981), on the basis of 178 observations made five days after launch. Significantly nonnormal pdfs are found, with kurtoses of 7.54 and 7.02, respectively; see Figure 11.10. For normal distributions, the kurtosis is 3. The mean square separations for

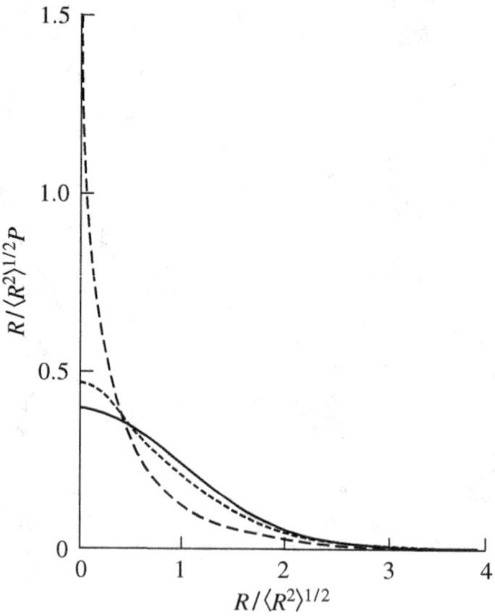

Figure 11.9 Theoretical and observed separation pdfs. The abcissa is $R/\langle R^2\rangle^{1/2}$ (here, $r/E\{r^2\}^{1/2}$). *Solid line*: relative diffusivity independent of R (here, r), according to Batchelor (1952). *Dashed line*: relative diffusivity proportional to $R^{4/3}$ (here, $r^{4/3}$), according to Richardson (1926). *Dotted line*: relative diffusivity inferred from observations of dye concentration in Lake Huron, according to Sullivan (1971).

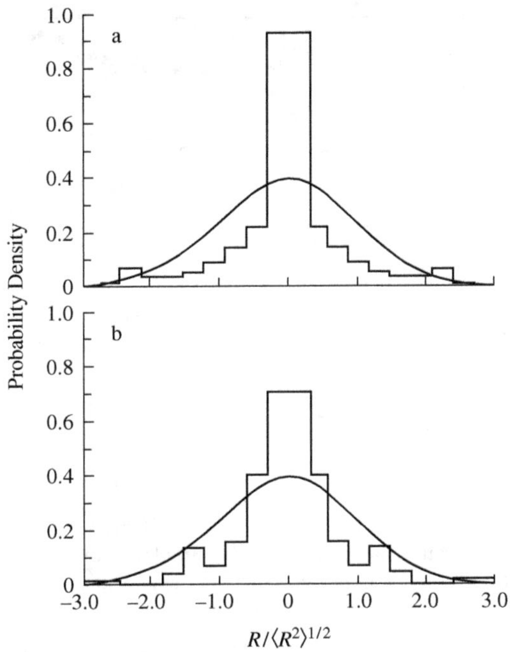

Figure 11.10 Sample pdfs for (a) zonal and (b) meridional components of separations of high-altitude ballons, five days after release in the southern hemisphere subtropics. The abcissa is $R/\langle R^2\rangle^{1/2}$ (here, $r/E\{r^2\}^{1/2}$). In each case, the normal pdf having the same mean and variance is included for reference. The kurtoses are 7.54 and 7.02, respectively; after Er-El and Peskin (1981).

the balloons exhibit exponential growth as in (11.37); see Figure 11.11. It is unfortunate that the kurtosis is not estimated at *two* times after launch.

Surface drifters deployed off the California coast by Davis (1985) are used to estimate separation pdfs; see Figure 11.12.

Pairs with initial separations in the range 16 km $< f <$ 30 km have, after four days, separations r closely consistent with a normal distribution for r_j. Those with initial separations in the range 4 km $< f <$ 16 km are more likely after four days to have smaller separations r than would be the case for normally distributed r_j. Davis (1985) attributes this finding to trapping in small-scale velocity convergences, or else to exponentially growing separations owing to large-scale deterministic shear. Davis (1985) also presents data purporting to show that η does not depend upon the separation r alone, but rather on r and the time $t-s$ elapsed since launch. However, it should be noted that what is shown is a dependence upon $\sqrt{E\{r^2\}}$ rather than on the conditional or observed value of r. This point is also discussed in Section 11.7.

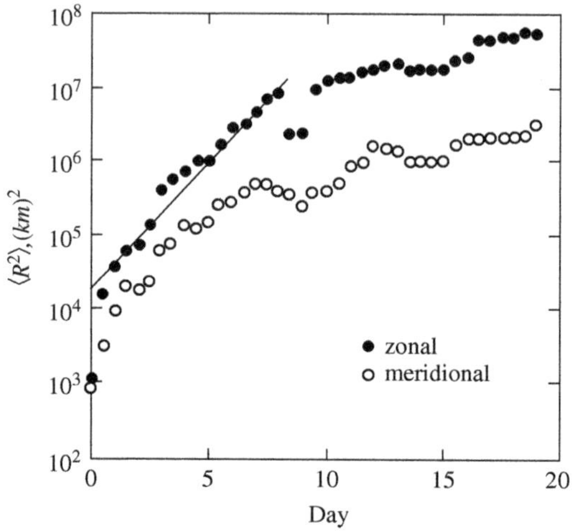

Figure 11.11 Separation variance $\langle R^2 \rangle$ (here, $E\{r^2\}$) in km^2 as a function of time in days from launch, from high-altitude balloons in the Southern Hemisphere tropics, the straight line indicating exponential growth; after Er-El and Peskin (1981).

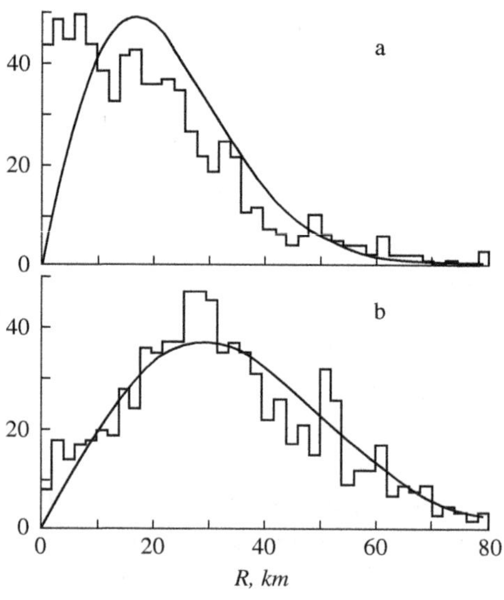

Figure 11.12 Histograms of separation of ocean surface drifters, four days after release off the California coast: (a) initial separations $4\,km < R < 16\,km$ (R is r here) and (b) initial separations $16\,km < R < 30\,km$. The histograms are based on bins 2 km wide. The smooth curves correspond to a normal distribution for $\mathbf{R}$ (here, $\mathbf{r}$); after Davis (1985).

LaCasce and Ohlmann (2003) clearly identify mean square separations growing exponentially in time, from observations of surface drifters in the Gulf of Mexico; see Figure 11.13. Such growth is consistent with (11.37). Moreover, the separation velocities for the drifters are correlated during this

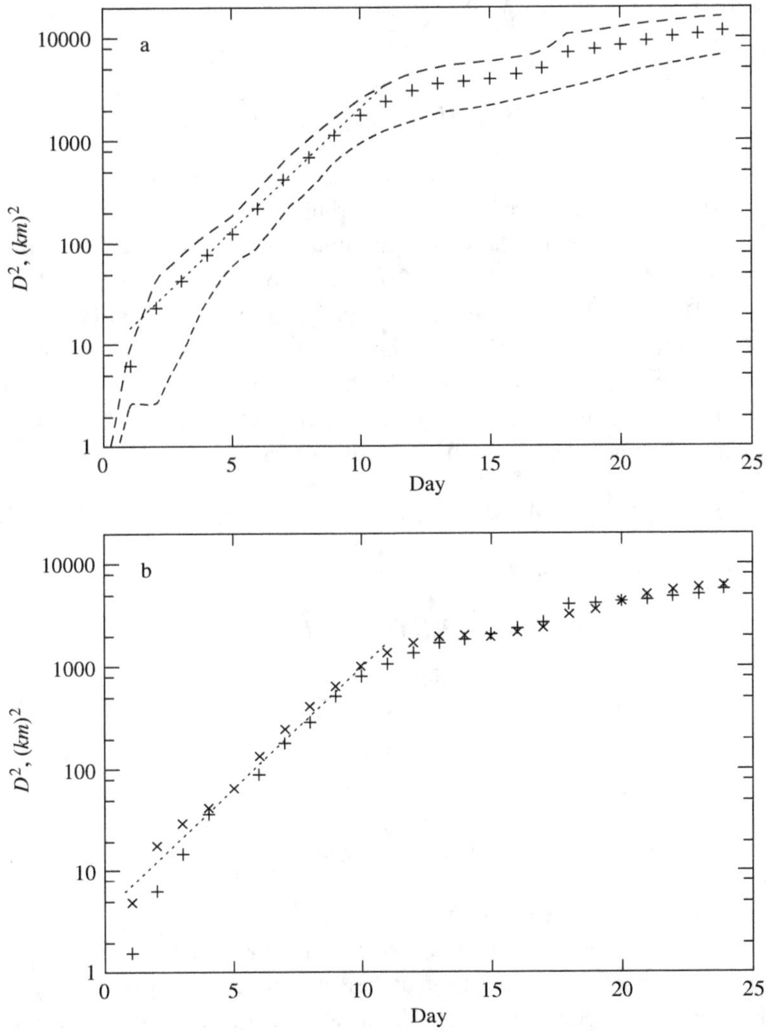

Figure 11.13 (a) The mean square separation D^2 (here, $E\{r^2\}$), in km^2, for surface drifters in the Gulf of Mexico *versus* time in days. The dashed lines indicate the 95% confidence limits and the straight line represents an exponential growth with a growth rate, determined by least squares, of 0.55 (day)$^{-1}$. (b) The mean square zonal and meridional separations *versus* time. After LaCasce and Ohlmann (2003).

mode of separation. On the other hand the kurtoses of separation are roughly constant during this mode, which is inconsistent with (11.37), but instead indicates a self-similar separation pdf, of which (11.32) is an example. Does the drifter separation owe to a large-scale random straining field, that is, to a nonlocal process, or rather to straining at the scale of the instantaneous separation value r, that is, to a local process?

11.6 Kinetic energy subranges

It will be seen in Chapters 12 and 13 that the variance spectrum of a conserved passive scalar concentration in isotropic turbulence is determined by the statistics of particle pair separation, which are in turn governed by the longitudinal diffusivity η. Again, Richardson (1926) infers from real atmospheric data that $\eta \propto r^{4/3}$. The high-altitude balloon data analyzed by Morel and Larcheveque (1974), and the Price data (McWilliams, et al., 1983) for deep ocean floats, indicate $\eta \propto r^2$. The form of η may also be inferred from a knowledge of the wavenumber spectrum of the Eulerian velocity field $u_j[x_k, t]$. For stationary isotropic turbulence the spectrum of kinetic energy is

$$E[k] = A(k) \int_0^\infty A(r) \mathcal{S}(kr) E\left\{u_j[x_i + r_i, t] u_j[x_i, t]\right\} dr, \tag{11.38}$$

where the low-pass filter $\mathcal{S}$ is

$$\begin{aligned} \mathcal{S}(kr) &= J_0(kr), & D &= 2; \\ \mathcal{S}(kr) &= \frac{\sin(kr)}{kr}, & D &= 3; \end{aligned} \tag{11.39}$$

see Figure 11.14.

The total kinetic energy is[2]

$$E_0 = \int_0^\infty E[k] dk. \tag{11.40}$$

It is assumed that this equilibrium spectrum of kinetic energy is maintained by a stationary isotropic source, at or around some low wavenumber l. The average source strength must be matched by the average dissipation rate ϵ for kinetic energy per unit mass, which rate has dimensions of $(\text{length})^2 \times$

[2] The appearances in equations of $E\{\ \}$ for ensemble average and $E[\ ]$ for energy spectrum should not be confusing if the braces and parentheses are noticed. These two uses of the letter E are conventional. Also, k denotes here a real-valued wavenumber magnitude rather than an integer-valued Cartesian index. It seems preferable to rely upon the reader to discern the meaning of a symbol from the context, rather than introducing yet more specialized symbols.

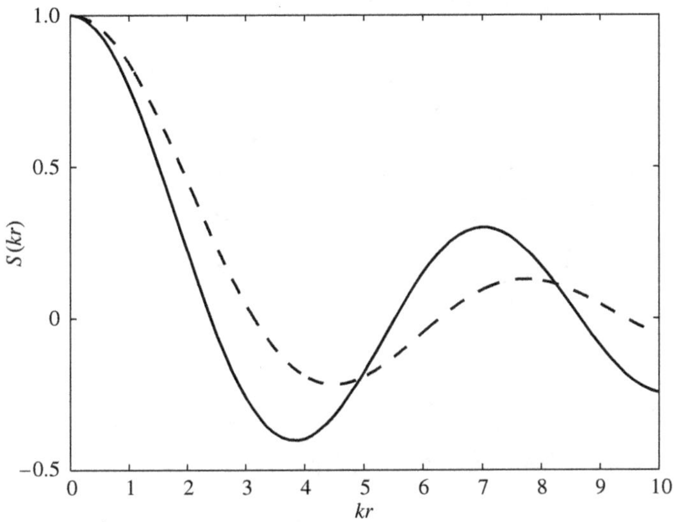

Figure 11.14 Low-pass filter $\mathcal{S}(kr)$ *versus* scaled wavenumber kr. *Solid line*: dimension $D=2$; *dashed line*: $D=3$.

(time)$^{-3}$. The dissipation rate is dominated by viscous dissipation at high wavenumbers. By assumption there are no sources or sinks at intermediate wavenumbers, so the energy spectrum in such an "inertial subrange" can only depend on ϵ and the local wavenumber k. Noting that $E[k]$ has the dimensions of (velocity)$^2 \times$ (length), dimensional analysis yields

$$E[k]=K_o\epsilon^{2/3}k^{-5/3}\,, \tag{11.41}$$

where K_o is the dimensionless Kolomogorov constant. For a comprehensive discussion, see Monin and Yaglom (1975). The inertial time-scale $\epsilon^{-1/3}k^{-2/3}$ exceeds the viscous time-scale $\nu^{-1}k^{-2}$ (where ν is the kinematic viscosity) if $k>k_\nu\equiv\epsilon^{1/4}\nu^{-3/4}$. Thus (11.41) can hold only if $l\ll k\ll k_\nu$. If $k_\nu\ll k$, then $E[k]$ must depend upon ϵ and ν, and dimensional analysis does not suffice. Observations (Grant et al., 1962; Pond et al., 1963; Gibson, 1963; Sreenivasan, 1995) confirm (11.41) at large Reynolds number, and also indicate a very rapid roll-off of $E[k]$ for $k\gg k_\nu$; see Figure 11.15. It suffices for our purposes to note only the fact of very rapid decay for $k\gg k_\nu$, which will be modeled here where necessary by the truncated spectral form

$$E[k]=K_o\epsilon^{2/3}k^{-5/3}\,, \qquad k<\psi k_\nu$$
$$E[k]=0\,, \qquad k>\psi k_\nu \tag{11.42}$$

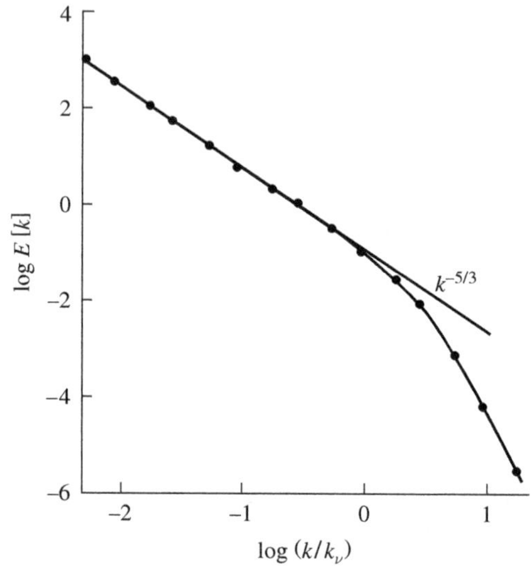

Figure 11.15 Wavenumber spectrum of kinetic energy $E[k]$ in a British Columbia tidal channel. The Kolomogorov wavenumber $k_\nu = (\epsilon/\nu^3)^{1/4}$ has the value $2.3 \times 10^3\ \mathrm{m}^{-1}$, given $\epsilon = 0.61 \times 10^{-4}\ \mathrm{m}^2\mathrm{s}^{-3}$ and $\nu = 1.28 \times 10^{-6}\ \mathrm{m}^2\mathrm{s}^{-1}$. The straight line has the slope $-5/3$; after Grant et al., (1962).

where ψ is some fraction. The observations of Grant et al. (1962), Pond et al. (1963) and Gibson (1963) all suggest that $\psi \cong 0.1$. At least one turbulence closure theory, the "Abridged Lagrangian History Direct Interaction Approximation" (Kraichnan, 1966a), not only conforms to the dimensionally obligatory (11.41) but is also in impressive agreement with the observations of the "viscous dissipation range"; see Figure 11.16.

A frequency spectrum $E[\omega]$ of either Eulerian velocity or Lagrangian velocity, at one point or for one release position, respectively, has the dimensions of $(\text{length})^2 \times (\text{time})^{-1}$, and so should be of the form $E[\omega] = c\epsilon\omega^{-2}$ in an inertial range $\omega_0 \ll \omega \ll \omega_\nu$, where ω_0 is the frequency of injection of momentum, $\omega_\nu \equiv (\epsilon/\nu)^{1/2}$ is the Kolmogorov frequency and c is a dimensionless constant. Lien and D'Asaro (2002) review theoretical studies and real observations of Lagrangian velocity; they find evidence of peaks in the acceleration spectrum $\omega^2 E[\omega]$ but typically the putative inertial range is too narrow for firm estimates for the proportionality constant c.

For inverse separations r^{-1} within the inertial subrange, where separation is controlled by eddies of wavenumber $k \sim r^{-1}$, the longitudinal diffusivity must on dimensional grounds be of the form

$$\eta = c\epsilon^{1/3} r^{4/3}, \tag{11.43}$$

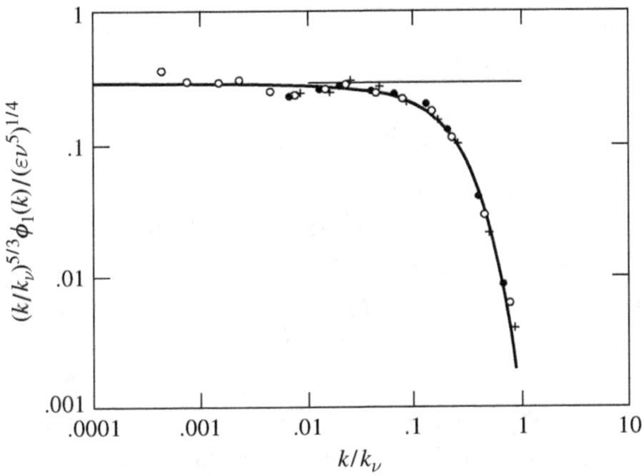

Figure 11.16 Nondimensional scaled energy spectrum $\phi_1(k)$ (here, $E[k]$) *versus* scaled wavenumber k/k_ν, according to the ALHDIA theory of Kraichnan (1966a).

where c is a dimensionless constant. Thus Richardson's observations (11.29) are consistent with the Kolmogorov energy subrange.

Kraichnan (1967) proposes alternative subranges in two-dimensional, stationary isotropic turbulence. The first is another "$-5/3$" range in which kinetic energy cascades not forward (that is, not from the low wavenumber l of the source to the wavenumbers above k_ν where viscous dissipation takes place), but rather in reverse from some intermediate wavenumber (possibly associated with some internal mechanism, such as baroclinic instability at the internal deformation radius) towards low wavenumbers. The direction of cascade is immaterial to the preceding analysis of particle separation statistics. It is necessary only to employ the form of $A(r)$ for $D=2$ when solving (11.23), (11.24) for the separation pdf.

Second, Kraichnan proposes a forward cascade of vorticity variance or "enstrophy." The total enstrophy Ω_0 is

$$\Omega_0 = E\left\{\left(\frac{\partial}{\partial x_1}u_2[x_1, x_2, t] - \frac{\partial}{\partial x_2}u_1[x_1, x_2, t]\right)^2\right\}. \tag{11.44}$$

At wavenumber k the enstrophy spectrum $\Omega(k)$ is

$$\Omega[k] = k^2 E[k]. \tag{11.45}$$

The total enstrophy is the integral of the enstrophy spectrum:

$$\Omega_0 = \int_0^\infty \Omega[k]dk. \tag{11.46}$$

The cascade rate λ for enstrophy has the dimensions of (time)$^{-3}$. Hence the energy spectrum must be

$$E[k] = K_r \lambda^{2/3} k^{-3}, \tag{11.47}$$

where K_r is a dimensionless constant that is appropriately named after Kraichnan. There is evidence of (11.47) in large-scale atmospheric circulation (Boer and Shepherd, 1983; see Figure 11.17). As might be expected, those large-scale data do not survive the stringent test for isotropy passed by smaller scale data supporting (11.41) (Gargett et al., 1984; Gargett, et al. 1985). Nor should the enstrophy inertial subrange extend to such high wavenumbers where, owing to nonhydrostatic effects, the flow is not approximately two-dimensional. Young et al. (1982) suggest that the highest wavenumber in the subrange may be $k \cong 10^{-3}$ m^{-1} in the ocean, or a shortest length-scale of one kilometer. The enstrophy spectrum corresponding to (11.47) is

$$\Omega[k] = K_r \lambda^{2/3} k^{-1}. \tag{11.48}$$

For inverse separations r^{-1} within the enstrophy inertial subrange, the longitudinal diffusivity must on dimensional grounds have the form

$$\eta = c\lambda^{1/3} r^2. \tag{11.49}$$

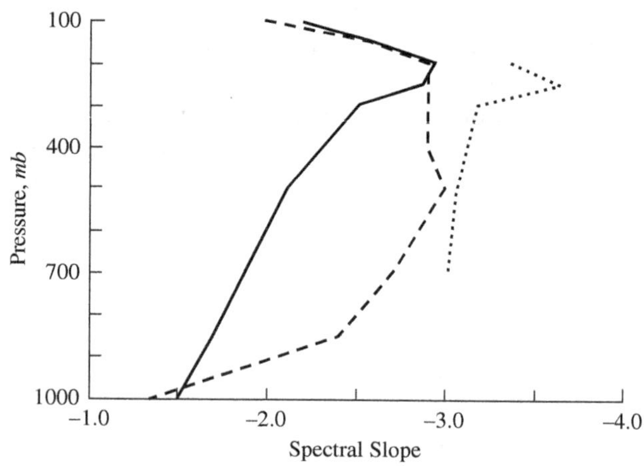

Figure 11.17 Slopes of straight-line fits to observations of the horizontal kinetic energy spectrum $\log E[n]$ *versus* $\log n$ in the atmosphere, for the zonal wavenumber n in the range $14 \leq n \leq 25$: *dotted line* from Baer (1972), *dashed line* from Chen and Wiin-Nielsen (1978), *solid line* from Boer and Shepherd (1983); after Boer and Shepherd (1983).

Thus the observations of Price (11.35) are consistent with Kraichnan's "−3" enstrophy cascading subrange. Note that subranges are characterized by the wavenumber exponent in the *energy* spectrum $E[k]$, thus the enstrophy cascading subrange is a "−3" subrange.

11.7 Kinetic energy spectra and structure functions

Consider again a pair of particles released at points a_i, b_j at time s. Their positions at time t are $x_n = X_n(a_i, s|t)$, $y_m = Y_m(b_j, s|t)$, and their separation is

$$r_l = y_l - x_l = X_l(b_j, s|t) - X_l(a_i, s|t); \tag{11.50}$$

see Figs. 11.1, 11.2. The velocity of separation is therefore

$$\frac{\partial r_l}{\partial t} = u_l(b_j, s|t) - u_l(a_i, s|t) = u_l[x_j + r_j, t] - u_l[x_j, t]. \tag{11.51}$$

Thus far, all expectations $E\{\ \}$ have been taken over the entire ensemble of velocity fields that define the random flow. Let a subensemble of velocity fields be defined, such that any member integrates to a particular value r_j for the pair separation vector at time t. The expectation of the squared magnitude of the separation velocity, taken over this subensemble, is

$$E\left\{\frac{\partial r_l}{\partial t}\frac{\partial r_l}{\partial t}\right\}\Bigg|_{r_j} = E\Big\{\big(u_l[x_j + r_j, t] - u_l[x_j, t]\big)\big(u_l[x_j + r_j, t] - u_l[x_j, t]\big)\Big\}. \tag{11.52}$$

The rhs of (11.52) is a purely Eulerian formula, since the value of the separation r_j is specified, that is, the value *conditions* the expectations on the lhs and rhs. It is straightforward to express (11.52) in terms of the kinetic energy spectrum (Kraichnan, 1966b; Bennett, 1984):

$$E\left\{\frac{\partial r_l}{\partial t}\frac{\partial r_l}{\partial t}\right\}\Bigg|_{r_j} = 2\int_0^\infty E[k]\big(1 - \mathcal{S}(kr)\big)dk, \tag{11.53}$$

where $\mathcal{S}(kr)$ is defined in (11.39). Thus, in stationary isotropic turbulence, the mean square separation velocity or *velocity structure function* at vector separation r_j is a function of only the scalar separation r. The high-pass filter acting on the energy spectrum in (11.53) has the limiting behavior

$$\begin{aligned} 1 - \mathcal{S}(kr) &\approx 1 + O\big((kr)^{-(D-1)/2}\big), \qquad kr \gg 1 \\ 1 - \mathcal{S}(kr) &\approx O\big((kr)^2\big), \qquad kr \ll 1; \end{aligned} \tag{11.54}$$

see Figure 11.18.

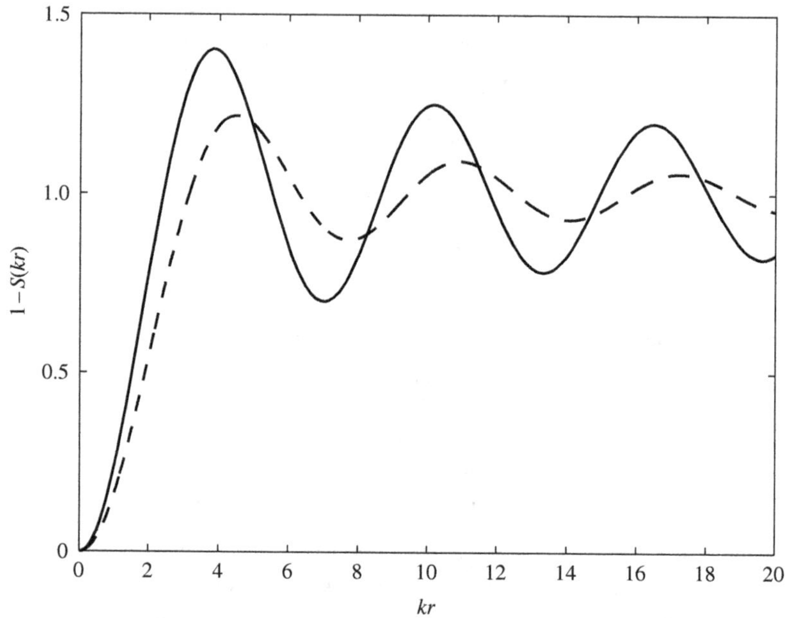

Figure 11.18 High-pass filter $1-\mathcal{S}(kr)$ *versus* scaled wavenumber kr. *Solid line*: dimension $D=2$; *dashed line*: $D=3$.

That is, the smaller eddies (larger k) send the ends of the lines on independent random walks (absolute dispersion), while the larger eddies (smaller k) elongate lines of length r by coherent stretching (relative dispersion). To elaborate, suppose there is a wavenumber subrange $l \ll k \ll h$ in which the energy spectrum has the form $E[k] \propto k^{-\alpha}$, and suppose that r^{-1} lies in this wavenumber range.

(i) If $\alpha \leq 1$, then the integral in (11.53) diverges as $k \to \infty$ (that is, until the very rapidly decaying viscous subrange is reached). Hence the structure function is dominated by the energy on the smallest scales in the range: dispersion is absolute rather than relative. Furthermore, (11.53) approximates to

$$E\left\{\frac{\partial r_l}{\partial t}\frac{\partial r_l}{\partial t}\right\}\Bigg|_{r_j} \approx 2E_0\,. \tag{11.55}$$

The two particles are moving independently.

(ii) If $1 < \alpha < 3$, then the integral in (11.53) converges as $k \to 0$ and as $k \to \infty$, so the structure function is dominated by the energy spectrum on wavenumbers of order r^{-1}. Dispersion is relative, and *local self-similarity* arguments, such as lead to the forms (11.42), (11.48) for the longitudinal diffusivity, are valid. Simple rescaling of (11.53) yields

$$E\left\{\frac{\partial r_l}{\partial t}\frac{\partial r_l}{\partial t}\right\}\bigg|_{r_j} \cong O\left(r^{\alpha-1}\right), \tag{11.56}$$

since $\mathcal{S}(kr) = O(1)$ for $kr = O(1)$.
Indeed, applying the local self-similarity arguments directly to the lhs of (11.52) yields (Morel and Larcheveque, 1974)

$$E\left\{\frac{\partial r_l}{\partial t}\frac{\partial r_l}{\partial t}\right\}\bigg|_{r_j} \cong r^{-1}E\left[r^{-1}\right], \tag{11.57}$$

since the expectation on the lhs is the kinetic energy of separation, while the spectrum on the rhs is the kinetic energy per unit wavenumber. The pure self-similar scaling formula (11.57) is consistent with (11.54) and with the assumption that $E[k] \propto k^{-\alpha}$. Recall that $\alpha = 5/3$ in the two- and three-dimensional energy cascading subranges (that is, $1 < \alpha < 3$) and so self-similarity arguments are justified. However, $\alpha = 3$ in the two-dimensional enstrophy cascading subrange, and so self-similarity arguments are only marginally justified.

(iii) If $\alpha \geq 3$, then the integral in (11.53) diverges as $k \to 0$ and the structure function is dominated by the largest energy-containing eddies. Dispersion is relative, and is controlled by *nonlocal* kinematics, that is, by wavenumbers of $o\left(r^{-1}\right)$. If $\alpha = 3$, the divergence is only logarithmic; there is a significant contribution to dispersion from eddies as small as r, the kinematics are only weakly nonlocal and no simple asymptotics apply to (11.53). If $\alpha \gg 3$, then the largest eddies completely dominate ($kr \ll 1$) and so

$$E\left\{\frac{\partial r_l}{\partial t}\frac{\partial r_l}{\partial t}\right\}\bigg|_{r_j} \cong c\Omega_0 r^2, \tag{11.58}$$

where c is a dimensionless constant and Ω_0 is the total enstrophy (11.46).

(iv) The local form (11.56), valid for $1 < \alpha < 3$, matches the nonlocal form (11.58) as $\alpha \to 3$ from below. It is therefore to be expected that (11.58) hold also for weakly nonlocal dynamics ($\alpha = 3$). Indeed, Lin (1972) proposes (11.58) as the self-similar form of the structure function for the enstrophy subrange ($\alpha = 3$), the total enstrophy Ω_0 being replaced by $\lambda^{2/3}$ where λ is the enstrophy cascade rate. However, the relationship (11.53) shows that the kinematics are weakly nonlocal in that subrange, and so self-similarity arguments are only marginally applicable.

11.8 Kinetic energy spectra and longitudinal diffusivities

The relative diffusivity defined in (11.26) may be expressed (Kraichnan, 1966b) as

$$\zeta(r,t|s) = E\left\{\frac{\partial}{\partial t}r^2\right\}\Bigg|_r$$

$$= 2E\left\{f_j\frac{\partial r_j}{\partial t}\right\}\Bigg|_r + 4\int_0^\infty E[k](1-\mathcal{S}(kr))\int_s^t R(k|t-w)dwdk\,. \tag{11.59}$$

In (11.59), f_j is the initial separation vector which may be random, while $R(k|t-w)$ is the normalized Lagrangian energy spectrum:

$$R(k|t-w) = \frac{Q(k|t-w)}{Q(k|0)}\,, \tag{11.60}$$

where

$$Q(k|t-w) = (2\pi)^{-D}\int \exp(ik_jr_j)E\left\{u_l[x_n+r_n,t]u_l(x_n,t|w)\right\}dV(r)\,, \tag{11.61}$$

and hence

$$Q(k|0) = A(k)^{-1}E(k)\,. \tag{11.62}$$

Note that R in (11.59) depends upon $t-w$; it is a normalized Lagrangian energy spectrum for stationary turbulence at time w, with labeling at time t.

The first term on the rhs of (11.59) vanishes if the direction of f_j, which is the separation vector at launch time s, is independent of the separation velocity $\partial r_j/\partial t$ at time t. Of course, the effective launch occurs once a high-altitude balloon or deep-ocean float reaches its equilibrium density surface. Thus f_j may have some correlation with $\partial r_j/\partial t$ at time s. In this regard, there is no clear distinction between "original" and "chance" pairs. In either case the correlation vanishes after several shear time scales $\left(\text{say, } (t-s) > 2\Omega_0^{-1/2}\right)$. Morel and Larcheveque (1974) find no significant differences in relative diffusivities for original and chance balloon pairs, and nor do LaCasce and Bower (2000) for float pairs.

Exercise 11.9 The longitudinal diffusivity $\eta(r, t|s)$ appearing in the Richardson–Kraichnan equation (11.23) is related via the integral (11.25) to the relative diffusivity $\zeta(r, t|s)$ defined in (11.26). Assuming no correlation between the separation velocity and the initial separation, show that (11.59) and (11.25) yield

$$\eta(r, t|s) = 4\int_0^\infty \int_s^t E[k]R(k|t-w)\mathcal{H}(kr)\,dw\,dk\,, \tag{11.63}$$

where

$$\mathcal{H}(kr) = r^{-D}\int_0^r \rho^{D-1}\big(1-\mathcal{S}(k\rho)\big)d\rho \tag{11.64}$$

is a high-pass filter arising from the geometry of isotropic turbulence:

$$\begin{aligned} &\mathcal{H}(kr) \to D^{-1}\,, && kr \gg 1,\\ &\mathcal{H}(kr) \propto O\big((kr)^2\big)\,, && kr \ll 1; \end{aligned} \tag{11.65}$$

see Figure 11.19. □

Assume again that $E[k] \propto k^{-\alpha}$, and in addition that $R(0|t-w) \neq 0$.

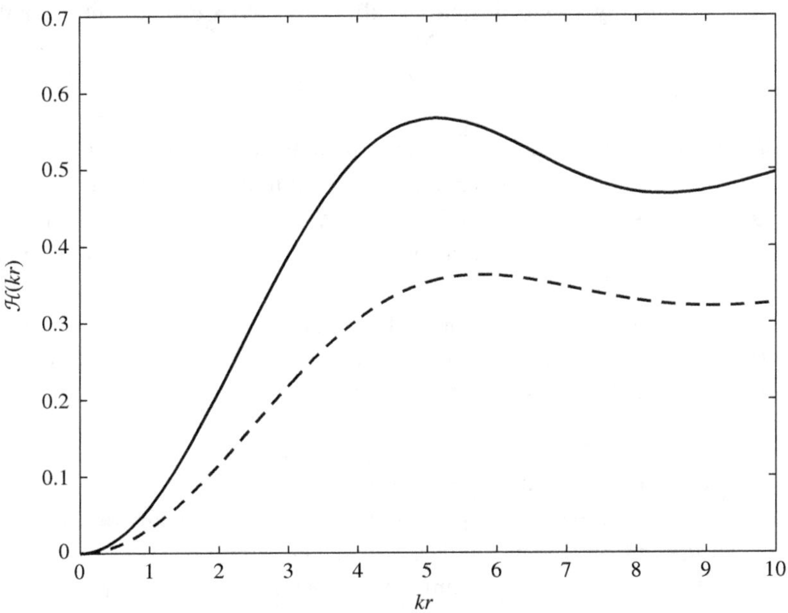

Figure 11.19 High-pass filter $\mathcal{H}(kr)$ *versus* scaled wavenumber kr. *Solid line*: dimension $D=2$; *dashed line*: $D=3$. Note the limits $\mathcal{H} \to D^{-1}$ as $kr \to \infty$.

(i) If $\alpha \leq 1$, then the wavenumber integral in (11.63) diverges as $k \to \infty$: dispersion is absolute rather than relative.
(ii) If $1 < \alpha < 3$, the integral converges as $k \to 0$ and as $k \to \infty$ (it may be assumed that R is bounded for large k), thus the kinematics are local. For sufficiently large t, the time integral in (11.63) becomes an integral time-scale which may be estimated (Kraichnan, 1966b) using local self-similarity arguments:

$$\int_s^t R(k|t-w)dw \sim E[k]^{-1/2}k^{-3/2} \propto k^{(\alpha-3)/2}. \qquad (11.66)$$

The wavenumber integral in (11.63) is dominated by $k \sim r^{-1}$, leading to

$$\eta \propto r^{(\alpha+1)/2}, \qquad (11.67)$$

in agreement with the direct scaling argument of Morel and Larcheveque (1974). Note that

$$\text{length} \sim r\,, \ \text{velocity} \sim \left(r^{-1}E[r^{-1}]\right)^{1/2}, \ \eta \sim \text{length} \times \text{velocity}. \qquad (11.68)$$

It is marginally sound, in the weakly nonlocal case $\alpha = 3$, to infer that $\eta \propto r^2$.

Exercise 11.10 Assuming a longitudinal diffusivity independent of time and having the form $\eta \propto r^{\beta}$, solve the initial value problem for the Richardson–Kraichnan equation (11.23), (11.24). Show that, asymptotically for large t, the solution for the separation pdf P has a self-similar form. The similarity variable is $r^{2-\beta}/(t-s)$, which is independent of the number D of space dimensions. The self-similar form for P does depend on D, however. Hint: the tables in Kamke (1959) and in Erdélyi et al. (1954) are helpful. Consider the limit as $\beta \to 2$. □

It must be concluded from Exercise 11.10 that a constant kurtosis is incompatible with an exponentially growing mean square separation. Thus the kurtosis, which LaCasce and Ohlmann (2003) estimate to be broadly unchanging in the range of exponentially growing mean square separation, must be passing through a local extremum in time rather than being in a constant state, to the extent that the two states are distinguishable in practice.

Exercise 11.11 Consider the "−5/3" energy spectrum (11.41), which characterizes directly or indirectly cascading energy subranges having dissipation rate ϵ. Show that on dimensional grounds

$$R(k|t-w)=U_1\left(\epsilon^{1/3}k^{2/3}(t-w)\right) , \tag{11.69}$$

where U_1 is some universal function, and hence

$$\eta \sim c_1 \epsilon^{1/3} r^{4/3} \tag{11.70}$$

as $(t-s)\epsilon^{1/3}r^{-2/3}\to\infty$, where c_1 is a dimensionless constant. Show also that the relative diffusivity is

$$\zeta = E\left\{\frac{\partial}{\partial t}r^2\right\}\bigg|_r \sim c_2\epsilon^{1/3}r^{4/3} , \tag{11.71}$$

where c_2 is another dimensionless constant. □

Seemingly substantial support for (11.71) has been obtained frequently since Richardson's pioneering study, over a very wide range of scales: $10\,\text{m} < r < 10^7\,\text{m}$! (Okubo, 1971); see Figure 11.20.

However, flow on the larger scales is hardly described as isotropic turbulence characterized by a well-defined ϵ. As Okubo (1971) points out, diagrams like Figure 11.20 can be misleading; they are not plots of ζ *versus* r, but rather E$\{\zeta\}$ it *versus* $\sqrt{E\{r^2\}}$. Thus all the "4/3" curve substantiates is a cubic time dependence: $E\{r^2\}\propto t^3$. The latter is also characteristic of particle pairs taking independent random walks in a shear flow (Bowden, 1965). In conclusion, it is doubtful that there is any genuine observational evidence in nature for (11.71), although laboratory measurements (Mory and Hopfinger, 1986) are suggestive.

Exercise 11.12 Consider a Eulerian velocity field with mean shear, $E\{u_i[x_j, t]\}=\gamma x_3\delta_{i1}$, where γ is a constant. Suppose that only a transverse velocity component is fluctuating: $E\{(u_1')^2\}=E\{(u_2')^2\}=0$, but $E\{(u_3')^2\}>0$. At time $t=0$, let two particles have separation $(f_1, 0, 0)$ which is sufficiently large in magnitude that their velocities are independent. Their subsequent motions may be regarded as independent random walks, and the kinematics of their separation may be modeled as

$$dr_1=\gamma r_3 dt, \quad dr_2=0, \quad dr_3=2K^{1/2}d\beta(t) , \tag{11.72}$$

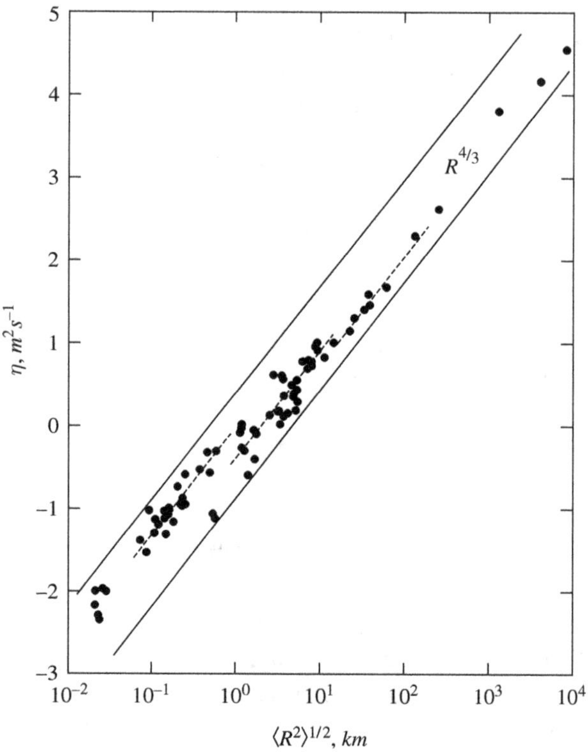

Figure 11.20 Relative diffusivity η (here, $E\{\zeta\}$) *versus* rms separation $\langle R^2\rangle^{1/2}$ (here, $E\{r^2\}^{1/2}$), inferred from observations at the ocean surface in various experiments. The straight lines have the slope +4/3, but this merely indicates that $\langle R^2\rangle \propto t^3$; after Okubo (1971).

where K is a constant diffusivity and $d\beta(t)$ is the Wiener process (e.g., Gardner, 1985). Show that $E\{r_3^2\} = 4Kt$, $E\{r_1 r_3\} = 2\gamma K t^2$, and so

$$E\{r_1^2\} = f_1^2 + \frac{4}{3}\gamma^2 K t^3 . \tag{11.73}$$

Note that this t^3 law is not a consequence of energy cascading turbulence. □

(iii) If $\alpha \geq 3$, the wavenumber integral diverges as $k \to 0$ and dispersion is controlled nonlocally. Weakly nonlocal kinematics ($\alpha = 3$) are characterized by the time-scale $\lambda^{-1/3}$ where λ is the enstrophy cascade rate, and so the Lagrangian correlation can only have the form

$$R(k|t-w) = U_2\big(\lambda^{1/3}(t-w)\big) , \tag{11.74}$$

where U_2 is some other universal function. The double integral in (11.63) then separates, yielding essentially the integral in (11.53) which is logarithmically divergent as $k \to 0$. The strongly nonlocal case needs no special consideration since, in that case ($\alpha \gg 3$) the Lagrangian correlation is again insensitive to k. Indeed, Kraichnan (1971) argues that the decorrelation time $T[k]$ at wavenumber k is given by the cumulative enstrophy in larger scales:

$$T[k]^{-2} = \int_0^k \Omega[l]dl = \int_0^k l^2 E[l]dl\,, \tag{11.75}$$

which converges rapidly ($\alpha \gg 3$) for large k to the total enstrophy Ω_0. Thus R is approximately of the form

$$R(k|t-w) \approx U_3\big(T^{-1}(t-w)\big) \tag{11.76}$$

for yet another universal function U_3. This leads to

$$\eta = c_3 T^{-1} W\big(T^{-1}(t-s)\big) r^2 \tag{11.77}$$

where c_3 is a dimensionless constant and

$$W(z) = \int_0^z U_3(z')dz'\,. \tag{11.78}$$

In particular $W(z) \to c_4$, where c_4 is another dimensionless constant. Thus the large-time longitudinal diffusivity is

$$\eta \approx c_5 T^{-1} r^2\,, \quad t-s \gg T\,. \tag{11.79}$$

The match between (11.79) and (11.67) for $\alpha = 3$ indicates that (11.77) is valid for both strongly and weakly nonlocal kinematics, with $\lambda^{-1/3}$ replacing T in the latter. Thus, mid-latitude observations of high-altitude balloons separating exponentially in time (Er-El and Peskin, 1981) are consistent with $\alpha \geq 3$.

Arguments analogous to the preceding, applied to weakly or strongly nonlocal kinematics, show that the relative diffusivity is

$$\zeta = E\left\{\frac{\partial}{\partial t} r^2\right\}\Bigg|_r = c_5 T^{-1} r^2\,, \tag{11.80}$$

with $T = \lambda^{-1/3}$ in the weakly nonlocal case. Thus the high-altitude balloon analyses of Morel and Larcheveque (1974) are also consistent with $\alpha \geq 3$.

Thus far, r has been a conditioned random variable. Averaging (11.80) over all r leads, for $\alpha \gg 1$, to

$$\frac{\partial}{\partial t} E\left\{r^2\right\} = c_5 T^{-1} E\left\{r^2\right\}, \quad t - s \gg T. \tag{11.81}$$

Averaging over the conditional separation in self-similar subranges leads nowhere. For example, in the energy subranges the averaged relative diffusivity is

$$\zeta = \frac{\partial}{\partial t} E\left\{r^2\right\} = c_6 \epsilon^{2/3} E\left\{r^{4/3}\right\}, \tag{11.82}$$

which unlike (11.81) is not a closed relationship for a moment of the separation r. On the other hand, dimensional analysis (Batchelor, 1952) leads directly to

$$\frac{\partial}{\partial t} E\{r^2\} = c_7 \epsilon t^2, \tag{11.83}$$

which implies the t^3 law for mean square separation, asymptotically for large t.

It is not clear that these asymptotic formulae for the longitudinal diffusivity η and the relative diffusivity ζ hold long before $\sqrt{E\{r^2\}}$ is much greater than the energy containing eddies, by which time dispersion has become absolute rather than relative. Nevertheless, (11.80) is well supported by atmospheric and oceanic data, while (11.83) is very well supported in synthetic turbulence having an energy range spanning many decades of wavenumber (Elliott and Majda, 1996).

12

Convective subranges of the scalar variance spectrum

12.1 Scalar covariance

Let C be the concentration per unit mass of a conserved, passive scalar. That is,

$$C(a_j, s|t) = C_I[a_j, s] + \int_s^t S(a_j, s|w)dw, \tag{12.1}$$

where $C_I[a_j, s] = C(a_j, s|s)$ is the concentration at time s, and $S(a_j, s|t)$ is the distributed source strength encountered at time t. Note that the dimension of S is (concentration) $\times$ (time)$^{-1}$.

Exercise 12.1 Show that

$$C[x_n, t] = C_I\big[X_j(x_n, t|s), s\big] + \int_s^t S\big[X_j(x_n, t|w), w\big]dw. \tag{12.2}$$

Now assume for simplicity alone that the initial concentration vanishes: $C_I = 0$. Assume that the source strength S is a random field, independent of the velocity field, and with vanishing Eulerian expectation:

$$E\big\{S[x_n, t]\big\} = 0. \tag{12.3}$$

Show that the mean concentration also vanishes:

$$E\big\{C[x_n, t]\big\} = 0, \tag{12.4}$$

and that the single-time, two-point Eulerian covariance of C is

$$\begin{aligned} E\big\{C[x_n, t]C[y_m, t]\big\} = \int_s^t \int_s^t \iint & E\big\{S[\alpha_k, v]S[\beta_l, w]\big\} \\ & \times E\left\{\delta\big(\alpha_k - X_k(x_n, t|v)\big)\delta\big(\beta_l - X_l(y_m, t|w)\big)\right\} \\ & dV(\alpha)dV(\beta)dvdw. \end{aligned} \tag{12.5}$$

Note that the joint pdf of the pair positions in (12.5) depends upon two observation times: v and w, as well as the common release time t.

Now assume that the random Eulerian source strength S is homogeneous in space and stationary in time, and moreover is uncorrelated in time:

$$E\{S[\alpha_k, v]S[\beta_l, w]\} = \Sigma[\alpha_k - \beta_k]\delta(v - w)\,, \tag{12.6}$$

where $\Sigma = \Sigma[r_k]$ is the as yet unspecified spatial covariance factor for the source, and $\delta = \delta(t)$ is as always the Dirac delta function. Note that the dimensions of Σ are (concentration)$^2 \times$ (time)$^{-1}$, since the dimension of $\delta(t)$ is (time)$^{-1}$

Hence show that

$$E\{C[x_n, t]C[y_m, t]\} = \int_s^t \iint \Sigma[a_k - \beta_k]P(x_n, y_m, t|\alpha_k, \beta_l, w) \times dV(\alpha)dV(\beta)dw\,. \tag{12.7}$$

□

Transforming α_k, β_k to centroid and separation coordinates, and integrating over the centroid, reduces (12.7) to

$$E\{C[x_n, t]C[y_m, t]\} = \int_s^t \int \Sigma[\rho_k]P(y_n - x_n, t|\rho_k, w)dV(\rho)dw\,, \tag{12.8}$$

since it also now assumed that the turbulent velocity field is statistically homogeneous, and so the marginal pdf for the separation $\rho_k = \beta_k - \alpha_k$ at time s is independent of the centroid $(x_n + y_n)/2$ at time t. Indeed, if the random source strength is statistically isotropic:

$$\Sigma[\rho_k] = \Sigma[\rho]\,, \tag{12.9}$$

where $\rho = \sqrt{\rho_k \rho_k}$, then so is the scalar concentration:

$$B[r, t] \equiv E\{C[x_n, t]C[x_m + r_m, t]\} = \int_s^t \int \Sigma[\rho]P(r, t|\rho, w)A(\rho)d\rho dw\,, \tag{12.10}$$

where $r_m = y_m - x_m$ and $r = \sqrt{r_n r_n}$. The pdf in (12.10) is the area average of the pdf in (12.8):

$$P(r, t|\rho, w) = A(\rho)^{-1} \int P(r_n, t|\rho_k, w)dA(\rho). \tag{12.11}$$

The area average is, as pointed out by Lundgren (1981), independent of the direction of the separation vector r_n when the velocity is statistically isotropic.

12.2 Reversibility

Consider now the Richardson–Kraichnan equation (11.23), subject to the initial condition (11.24), for the approximate evolution of the separation pdf $P(f, s|r, t)$. Is the solution suitable for the evaluation of the isotropic, stationary, single-time, two-point concentration covariance $B[r, t]$ via (12.10)? The labeling time t in (12.10) exceeds the running or observation time w, yet (11.23), (11.24) may only be integrated for an observation time t exceeding the labeling time s, since the longitudinal diffusivity η is assumed to be nonnegative: see (11.25), (11.26).

Exercise 12.2 A consequence of assuming incompressible flow is that the arguments of the joint pdf for the positions of a pair of particles are reversible: see (11.5). Show therefore that

$$P(f, s|r, t) = P(r, t|f, s)\,. \tag{12.12}$$

Hence (12.10) becomes

$$B[r, t] = \int_s^t \int \Sigma[\rho]P(\rho, w|r, t)A(\rho)d\rho dw\,, \tag{12.13}$$

which permits the use of solutions of (11.23), (11.24). □

Exercise 12.3 Show that

$$\frac{\partial}{\partial t}B[r, t] = A(r)^{-1}\frac{\partial}{\partial r}\left\{A(r)\eta(r, t|s)\frac{\partial}{\partial r}(B[r, t])\right\} + \Sigma[r]\,. \tag{12.14}$$

What is a suitable initial condition? What is the solution if (i) $\eta \equiv 0$; (ii) η is a nonvanishing constant? □

12.3 Power spectra

The one-dimensional wavenumber spectrum of the Eulerian scalar concentration C is defined by

$$F[k, t] = \int dA(k) \int dV(r) \exp(ik_n r_n) E\{C[x_n, t]C[x_n + r_n, t]\}\,, \tag{12.15}$$

where $k = \sqrt{k_j k_j}$. It has been assumed that $E\{C[x_n, t]\} = 0$, and that C is statistically isotropic, so the covariance of C depends only upon $r = \sqrt{r_j r_j}$ and t. Thus (12.10) and (12.15) reduce to

$$F[k, t] = A(k) \int_0^\infty A(r)\mathcal{S}(kr)B[r, t]dr \tag{12.16}$$

where, again,

$$\begin{aligned} A(r) &= 2\pi r\,, \quad D=2 \\ A(r) &= 4\pi r^2\,, \quad D=3 \\ \mathcal{S}(kr) &= J_0(kr)\,, \quad D=2 \\ \mathcal{S}(kr) &= \frac{\sin(kr)}{kr}\,, \quad D=3; \end{aligned} \tag{12.17}$$

see Fig 11.14. The total variance is obtained by integrating the one-dimensional variance spectrum over all wavenumbers:

$$E\left\{C[x_n, t]^2\right\} = (2\pi)^{-D} \int_0^\infty F[k, t]dk\,. \tag{12.18}$$

Note that, consistent with the assumed homogeneity of C, the rhs of (12.18) is independent of x_n. It remains to substitute the scalar covariance, expressed by (12.10) in terms of the separation pdf, into (12.15):

$$F[k, t] = A(k) \int_0^\infty A(r)\mathcal{S}(kr) \int_0^\infty \Sigma[\rho]A(\rho) \int_s^t P(\rho, w|r, t)dwd\rho dr\,. \tag{12.19}$$

The time rate of change of scalar variance at wavenumber k is therefore

$$\frac{\partial}{\partial t}F[k, t] = \mathcal{J}[k, t]A(k) + \int_0^\infty A(r)\mathcal{S}(kr)\Sigma[r]dr\,, \tag{12.20}$$

where $\mathcal{J}[k, t]$ is the rhs of (12.19) with P replaced by $\partial P/\partial t$. The second term on the rhs of (12.20) is the source of scalar variance at wavenumber k.

Exercise 12.4 Assume for example that

$$\Sigma[r] = \chi\mathcal{S}(lr)\,, \tag{12.21}$$

where l is some wavenumber, and χ is the variance of the source strength:

$$\chi = \Sigma[0]\,. \tag{12.22}$$

Note that $\mathcal{S}$, which is given by (12.17), is dimensionless so the dimensions of χ are (concentration)^2 $\times$ (time)$^{-1}$. Show that

$$\frac{\partial}{\partial t}F[k, t] = \mathcal{J}[k, t] + (2\pi)^D\chi\delta(k-l)\,. \tag{12.23}$$

The source is isolated to a single wavenumber $k=l$. The term $\mathcal{J}[k, t]$ represents turbulent transfer of scalar variance to wavenumber k from other wavenumbers. Clearly, the development of an evolution equation for the separation pdf $P(\rho, w|r, t)$, such as (11.23), is equivalent to developing a model for the spectral transfer rate $\mathcal{J}[k, t]$.

Show also that

$$\int_0^\infty \mathcal{J}[k, t]dk = 0, \tag{12.24}$$

and hence

$$\frac{\partial}{\partial t} E\left\{C[x_n, t]^2\right\} = \chi. \tag{12.25}$$

That is, scalar variance is conserved in the absence of a source. Otherwise, the solution of (12.25) is unbounded for large t, as is to be expected in the absence of molecular diffusion of the scalar. □

12.4 Enstrophy inertia convective subrange

The enstrophy cascading inertial subrange of stationary, isotropic two-dimensional turbulence is characterized by the enstrophy cascade rate λ, the time-scale $T = \lambda^{-1/3}$ and the Eulerian kinetic energy spectrum

$$E[k] = K_r \lambda^{2/3} k^{-3}, \tag{12.26}$$

where K_r is the dimensionless Kraichnan constant. The longitudinal diffusivity has the large-time asymptote

$$\eta \sim c\lambda^{1/3} r^2, \tag{12.27}$$

where c is a dimensionless constant and r is the particle separation. The corresponding solution of the Richardson–Kraichnan equation (11.23) and initial condition (11.24) for the pdf of r at time t given a separation f at time s, for $D = 2$, is a log normal distribution (11.36) (Lundgren, 1981). The large-time limit $F[k]$ of the variance spectrum $F[k, t]$ of the Eulerian scalar concentration $C[x_j, t]$ may be explicitly calculated using the general relation (12.19), the specific form (12.21) for the spatial covariance $\Sigma[r]$ of the white noise source $S[x_j, t]$ as characterized by (12.6), and the log normal form (11.36) of the separation pdf $P(f, s|r, t)$. It is convenient to interchange orders of integration and then use the following result for P:

$$\lim_{t\to\infty} \int_s^t P(\rho, w|r, t)dw = c\lambda^{-1/3} r^{-2}, \quad \rho < r$$

$$\lim_{t\to\infty} \int_s^t P(\rho, w|r, t)dw = c\lambda^{-1/3} \rho^{-2}, \quad r < \rho, \tag{12.28}$$

where c is a dimensionless constant. Note that the time integrated pdf in (12.28) is not normalized over r; this is to be expected since the pdf in the integrand is normalized over r for each w. However, given (12.28) the

Fourier integrals in (12.19) are absolutely convergent and so interchanging orders of integration is justified. Combining (12.28) with (12.19) and (12.21) yields, finally,

$$F[k] = c\chi\lambda^{-1/3} k l^{-2}, \quad k < l, \tag{12.29}$$

$$F[k] = c\chi\lambda^{-1/3} k^{-1}, \quad k > l, \tag{12.30}$$

where c is a positive dimensionless constant. This spectral shape, $O(k^{-1})$ for large k, is neither red nor blue; every wavenumber decade above l makes the same contribution to the total scalar variance. The total is therefore infinite, as might be expected, given that there is a statistically stationary source of variance and that scalar diffusion is being ignored: see (12.25). It must be conceded that (12.30) could have been deduced by dimensional analysis alone, without having to determine the separation pdf or having to evaluate the integral (12.19). See below for further consideration of this point.

Exercise 12.5 Derive (12.28). It is easiest to apply the Laplace transform to (11.23) and (11.24) with transform variable q, and then simply to find the transformed solution as a function of r, for $q=0$. Then derive (12.29) and (12.30). To this end, define $G[k] \equiv k^{-1}F[k]$, and show that

$$\frac{\partial}{\partial k}\left(k^3 \frac{\partial}{\partial k} G[k]\right) = -c\chi\lambda^{-1/3}\delta(k-l), \tag{12.31}$$

where c is a positive dimensionless constant. □

12.5 Energy inertia convective subrange

The energy cascading inertial subrange of two- and three-dimensional stationary isotropic turbulence is characterized by the energy cascade rate ϵ, and the Eulerian kinetic energy spectrum

$$E[k] = K_o \epsilon^{2/3} k^{-5/3}, \tag{12.32}$$

where K_o is the dimensionless Kolmogorov constant. The longitudinal diffusivity has the large-time asymptote

$$\eta \sim c\epsilon^{1/3} r^{4/3}, \tag{12.33}$$

and for $D=3$ the corresponding solution of the Richardson–Kraichnan equation (11.23), (11.24) is given by (11.30). It suffices for the evaluation of the limiting scalar variance spectrum $F[k]$, using (12.19), to know that

$$\lim_{t\to\infty}\int_s^t P(\rho, w|r, t)dw = c\epsilon^{-1/3}r^{-7/3}, \quad \rho<r;$$

$$\lim_{t\to\infty}\int_s^t P(\rho, w|r, t)dw = c\epsilon^{-1/3}\rho^{-7/3}, \quad r<\rho. \tag{12.34}$$

The manipulations that yield the large-time asymptotic scalar variance spectrum $F[k]$ in the enstrophy inertia convective subrange are inadmissible here, owing not least to the Fourier integrals no longer being absolutely convergent. A more circumspect approach is effective, starting from the general relationship (12.13) and the model source covariance (12.21). First, the asymptotic scalar covariance is calculated as

$$B[r] = \lim_{t\to\infty} B[r, t] = 4\pi\chi\int_0^\infty \mathcal{S}(l\rho)\rho^2 \lim_{t\to\infty}\int_s^t P(\rho, w|r, t)dwd\rho, \tag{12.35}$$

where l is the injection wavenumber of the source. Recall that $D=3$, and hence $A(\rho)=4\pi\rho^2$. The interest is in separations $r\ll l^{-1}$, thus

$$B[r]\cong c\chi\epsilon^{-1/3}\big(l^{-2/3}-c'r^{2/3}\big) \tag{12.36}$$

where c and c' are dimensionless constants. Note that the total scalar variance is found to have a finite asymptotic value:

$$B[0]\cong c\chi\epsilon^{-1/3}l^{-2/3}, \tag{12.37}$$

even though scalar dissipation has been explicitly neglected (it is implicitly present in the scalar cascade rate χ). However, it will be seen that the total scalar dissipation rate has an infinite value rather than the correct value χ. It follows immediately from (12.36) that the asymptotic structure function for the scalar is

$$\lim_{t\to\infty} E\left\{\big(C[x_j+r_j, t]-C[x_j, t]\big)^2\right\} = 2\big(B[0]-B[r]\big)\cong c\chi\epsilon^{-1/3}r^{2/3}, \tag{12.38}$$

for $r\ll l^{-1}$. But

$$B[0]-B[r] = (2\pi)^{-3}\int_0^\infty F[k]\big(1-\mathcal{S}(kr)\big)dk, \tag{12.39}$$

so it follows immediately that

$$F[k]\cong c\chi\epsilon^{-1/3}k^{-5/3}. \tag{12.40}$$

Note that, with this solution, the integral in (12.39) is convergent, both as $k\to 0$ and as $k\to\infty$. There is a wealth of atmospheric and oceanic data supporting (12.40); see Figure 12.1 (Champagne et al., 1977) and Figure 12.2 (Gargett, 1985).

The result (12.40) was originally obtained using dimensional arguments alone by Obhukov (1949), Corrsin (1951) and Batchelor (1959). Detailed calculations as above would seem unjustified. Yet the success of these detailed calculations supports Richardson's "4/3" law (11.29). Batchelor (1952) argued, to the contrary, that the relative diffusivity ζ and hence the longitudinal diffusivity η should both be independent of r; on dimensional grounds this implies

$$\eta = c\epsilon(t-s)^2 , \tag{12.41}$$

and it follows that the diffusivity tensor η_{ij} should be simply $\eta\delta_{ij}$. Then (11.18) is readily solved for the pdf of the separation vector r_i. The result is an uncorrelated multivariate normal distribution with vanishing mean, and variance

$$E\{r^2\} = c\epsilon(t-s)^3 , \tag{12.42}$$

which is well supported by numerical simulations of stationary and isotropic turbulence (Elliot and Majda, 1996). Combining this result with the general

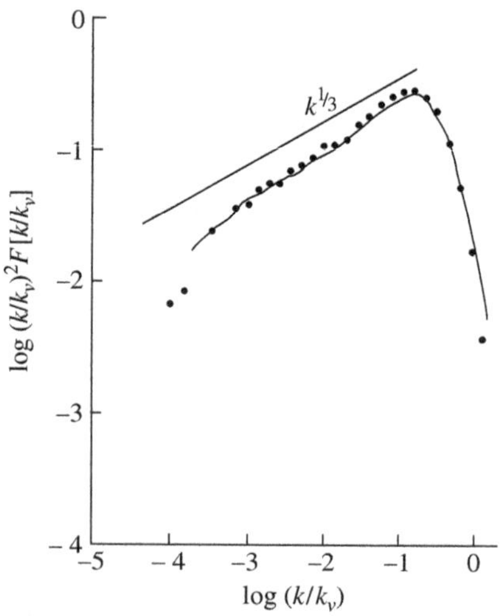

Figure 12.1 Temperature variance dissipation spectrum $(k/k_\nu)^2 F[k/k_\nu]$ *versus* k/k_ν, in an atmospheric boundary layer in Minnesota; after Champagne et al. (1977).

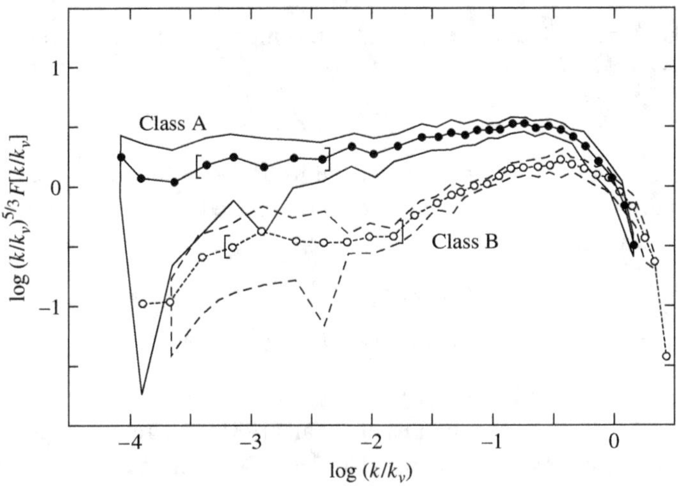

Figure 12.2 Class average, 5/3-moment temperature spectra $(k/k_\nu)^{5/3}F[k/k_\nu]$ in a British Columbia fjord *versus* k/k_ν. *Class A* data pass stringent tests for statistical isotropy and have high signal-to-noise ratios. *Class B* data depart from isotropy and have a lower signal-to-noise ratio. The envelopes indicate the variance within the classes. The approximately level segments between the brackets indicate $F[k] \propto k^{5/3}$; after Gargett (1985).

relation (12.19) and the model source covariance (12.21) leads to the asymptotic scalar variance spectrum

$$F[k] = c\chi\epsilon^{-1/3} l^{-2/3} \delta(k-l), \tag{12.43}$$

indicating no cascade of scalar variance. Note that $\delta(k-l)$ has the same dimensions as k^{-1}, and hence (12.43) is *dimensionally* correct.

12.6 Viscous convective subrange

Molecular diffusion of the scalar concentration may be negligible for wavenumbers far higher than the Kolmogorov wavenumber $k_\nu = \epsilon^{1/4}\nu^{-3/4}$ which is the upper bound for the energy cascading inertial subranges. The longitudinal diffusivity may be calculated from (11.63), (11.60) and (11.41) so long as the wavenumber integration is artificially cut off above $k = \psi k_\nu$. Recall that ψ is observed to be about 0.1, and that the observed energy spectra decay much faster than (11.40) above this cutoff. The dimensionless Lagrangian spectrum in (11.63) must have the self-similar form (11.66).

Since the case of interest is $r \ll k_\nu^{-1}$, the high-pass filter is approximately $\mathcal{H}(kr) = O\big((kr)^2\big)$, and so

$$\eta \cong c\tau^{-1} r^2 \,, \tag{12.44}$$

where c is a constant; $\tau = \epsilon^{-1/2}\nu^{1/2}$ is the Kolmogorov or advective time-scale at wavenumber k_ν. It may be immediately inferred that the asymptotic scalar variance spectrum is

$$F[k] = c\chi\tau k l^{-2} \,, \quad k < l \,; \tag{12.45}$$

$$F[k] = c\chi\tau k^{-1} \,, \quad k > l \,. \tag{12.46}$$

This result was originally obtained by Batchelor (1959), again by Kraichnan (1974), and also by Lesieur et al. (1981). There is no clear evidence for (12.46) in atmospheric or oceanic data. Convective subranges can only exist when the scalar diffusion rate κk^2 (κ is the scalar diffusivity) is much smaller than the rms strain rate τ^{-1} (see Exercise 12.6), that is, $k \ll k_B$ where $k_B = k_\nu P_r^{1/2}$ is the Batchelor wavenumber and $P_r = \nu/\kappa$ is the Prandtl number. Thus, viscous convective subranges require $k_\nu \ll k \ll k_\nu P_r^{1/2}$, or $1 \ll P_r^{1/2}$. For air, $P_r^{1/2} = 0.85$, while for water $P_r^{1/2} = 2.6$. However, several of the scalar spectra which Gargett (1985) reports show "$-5/3$" inertia convective subranges that flatten out before rolling off above k_B.

Exercise 12.6 In three-dimensional isotropic turbulence, the total enstrophy Ω_0 is the variance of the rate-of-strain tensor:

$$\Omega_0 = E\left\{ \frac{\partial u_i}{\partial x_j} \frac{\partial u_i}{\partial x_j} \right\} . \tag{12.47}$$

Verify (11.46), where the one-dimensional or directionless enstrophy spectrum $\Omega[k]$ is defined by (11.45). Note that the rhs of (12.47) is independent of position. If the isotropic turbulence is an energy cascade, show that Ω_0 is related to the Kolmogorov time-scale τ by

$$\Omega_0 = c\tau^{-2} \,, \tag{12.48}$$

where c is a constant. Is Ω_0 as well defined for an enstrophy cascade? □

12.7 Transition

The energy spectrum $E[k]$ has a well-defined transition from the (energy) inertial subrange to the viscous subrange at $k \cong 0.1\, k_\nu$; see Figure 11.15. Thus there should be a transition in the scalar variance spectrum from the "$-5/3$" inertia convective subrange to the "-1" viscous convective subrange, also at $k = 0.1k_\nu$. Yet observations (e.g., Gargett, 1985) show well-defined scalar transitions at $k = 0.01k_\nu$. Consider therefore the limiting form for the longitudinal diffusivity η as $(t-s) \to \infty$. The effective range of integration for k in (11.63) is, again, $0 < k < 0.1k_\nu$; in any case such a truncation can only promote the inertia convective subrange form for η (that is, possibly pushing the scalar transition towards wavenumbers higher than $0.1k_\nu$). Then, scaling as before,

$$\eta = c\epsilon^{1/3} \int_0^{\psi k_\nu} k^{-7/3} \mathcal{H}(kr) dk \tag{12.49}$$

where c is a constant. The integral over k may be evaluated numerically; see Figure 12.3.

There is a well-defined transition for η at a separation r almost a decade above than the cutoff length $(\psi k_\nu)^{-1}$. This indicates a scalar transition at a wavenumber almost a decade below ψk_ν. The source of this numerical factor of about 6 is the slow rise of the high-pass filter $\mathcal{H}(kr)$, from $(kr)^2/30$ for $kr \ll 1$ to $D^{-1} = 1/3$ as $kr \to \infty$; see Figure 11.19. Note that the first maximum of $\mathcal{H}$ occurs at $kr \cong 6$. The shape of $\mathcal{H}$ is purely a consequence of the geometry of isotropic turbulence.

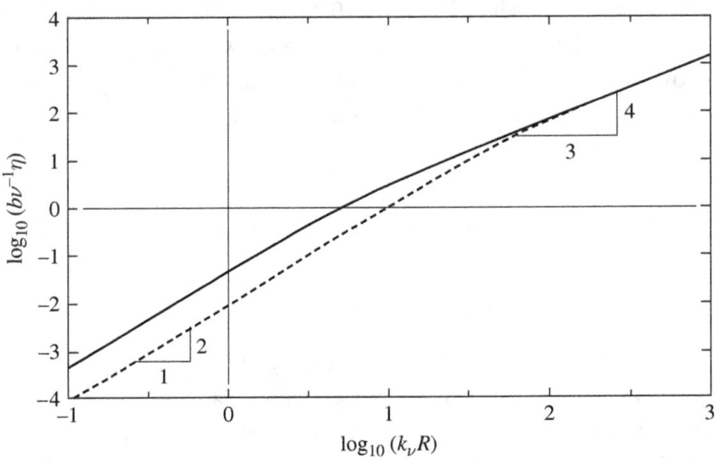

Figure 12.3 Large-time longitudinal diffusivity η *versus* scaled separation $k_\nu R$ (here, $k_\nu r$), according to the truncated model (11.42) for the energy spectrum. *Solid line*: for $\psi = 1$, the transition from the R^2 range to the $R^{4/3}$ range occurs at $R \approx 6k_\nu^{-1}$. *Dashed line*: for $\psi = 0.1$, the transition is at $R \approx 60k_\nu^{-1}$. That is, the transition occurs at a separation close to an order of magnitude larger than the length-scale of maximum dissipation; after Bennett (1987). The generic constant b (here, the generic c) in the scaling of η is dimensionless.

If the Prandtl number $P_r = \nu/\kappa$ is not large, then the viscous convective subrange should not be well defined. The inertia convective subrange should have a smooth transition to the rapidly decaying viscous diffusive subrange described in the next chapter. However, a clear bump in the scalar spectrum is commonly observed for $k = O(k_\nu)$; see Champagne et al. (1977) and Williams and Paulson (1977). This bump may be explained in terms of the latent viscous convective subrange that exists, owing to the tendency of η to convert from the "4/3" law to the "2" law for $r \gg k_\nu^{-1}$. Similar conclusions are reached by Hill (1978), who calculates spectra using several models for the spectral transfer rate. The models have several disposable parameters, in addition to ψ which is observed. The integral (12.49) has only ψ.

12.8 Relative dispersion and plankton patchiness

Consider again plankton concentration C, subject to advection by incompressible planar flow u_k, and growing along the particle path $x_j = X_j(a_k, s|t)$ according to the Lagrangian logistic model (10.114) where $r = r[x_j, t]$ is a field of growth rate of either sign, owing for example to nutrients or zooplankton grazing; C_p is a carrying capacity. The nonlinear transformation (10.115) leads to the simple conservation law (10.116) for D, with a simple source r. Bennett and Denman (1985) suppose that the growth rate field r is conserved by the flow and is source-free, such as an inexaustible supply of nutrients or grazing by a stable community of zooplankton:

$$\frac{\partial}{\partial t} r = 0 . \tag{12.50}$$

Assume without loss of generality that the initial transformed concentration D vanishes. It follows easily that

$$D[x_n, t] = r_I[X_m(x_n, t|0)]t , \tag{12.51}$$

where $r_I = r_I[x_m]$ is the growth rate at time $t = 0$. Assume further, as in Section 12.3, that r_I is statistically isotropic, with covariance $R^2 J_0(k_0 r)$. That is, the variance is R^2 and the variance spectrum is sharply peaked at some wavenumber k_0. If the velocity field in the wavenumber range $k_0 < k < k_{\max}$ is stationary isotropic turbulence cascading enstrophy from low to high wavenumbers at the rate λ as described in Section 11.6, then it follows from the log normal separation pdf of Section 11.4 that the time-dependent variance spectrum of D is

$$\widetilde{\Delta}[k, t] = t^2 R^2 (2\pi\sigma)^{-1/2} k^{-1} \exp\left[-\left(\ln(k/k_0) - \sigma\right)^2 (2\sigma)^{-1} \right], \tag{12.52}$$

where $\sigma = a\lambda^{1/3} t$ and a is a positive constant.

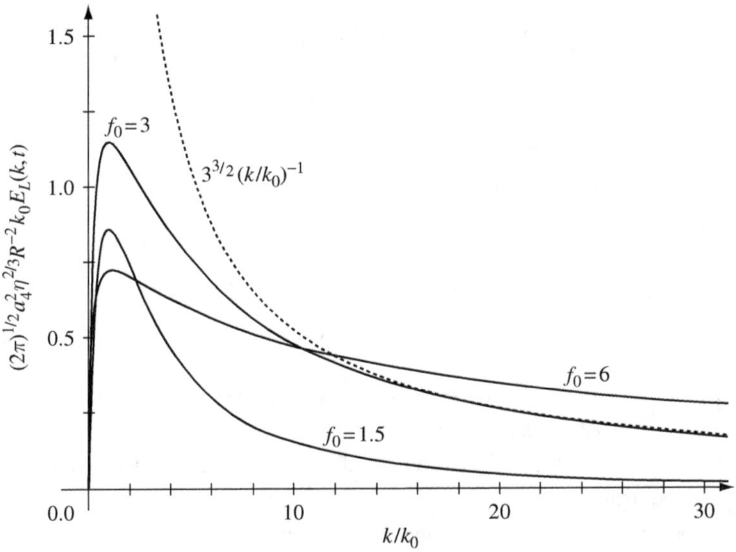

Figure 12.4 *Solid lines*: scaled variance spectra $(2\pi)^{1/2}a_4\eta^{2/3}R^{-2}k_0E_L(k,t)$ (here, $(2\pi)^{1/2}a\lambda^{2/3}r^{-2}k_0\widetilde{\Delta}[k,t]$) of advected growth rate *versus* (k/k_0) for f_0 (here, for $\sigma_0 = a\lambda^{1/3}t_0) = 3/2, 3, 6$. The wavenumber k_0 is the peak in the variance spectrum of imposed growth rates. *Broken line*: $f_0^{3/2}(k/k_0)^{-1}$ *versus* k/k_0 for the case $f_0=3$. After Bennett and Denman (1985).

Notes:

(i) $\widetilde{\Delta}$ has a solitary maximum in wavenumber at $k=k_0$, and the maximum value is

$$\widetilde{\Delta}[k_0, t] = t^2R^2(2\pi\sigma)^{-1/2}k_0^{-1}\exp(-\sigma/2)\,; \tag{12.53}$$

(ii) as a function of time, the wavenumber maximum has a solitary maximum at $t=t_0=3a^{-1}\lambda^{-1/3}$, and the maximum is

$$\widetilde{\Delta}[k_0, t_0] = t_0^2R^2(6\pi)^{-1/2}\exp(-3/2)\,, \tag{12.54}$$

thus for $t>t_0$ the new D variance created by r at wavenumber k_0 cascades to higher wavenumbers at a rate sufficient to erode the maximum (12.54);

(iii) the total variance of D is

$$\mathrm{var}\{D\}(t) = \int_0^\infty \widetilde{\Delta}[k,t]dk = t^2R^2\,, \tag{12.55}$$

which grows quadratically in time, as there is no explicit dissipation of D;

(iv) the first moment of $\widetilde{\Delta}$ about $k=0$ is

$$\overline{k_1}(t)=t^{-2}R^{-2}\int_0^{\infty} k\widetilde{\Delta}[k,t]dk=k_0\exp(3\sigma/2)\,, \tag{12.56}$$

which grows exponentially in time;

(v) there is a wavenumber range given by

$$\sigma-(2\sigma)^{1/2}<<\ln(k/k_0)<<\sigma+(2\sigma)^{1/2} \tag{12.57}$$

in which (12.52) has an approximate power-law behavior:

$$\widetilde{\Delta}[k,t]\cong t^2R^2(2\pi\sigma)^{-1/2}k^{-1}\,, \tag{12.58}$$

the center of the range being at $k=k_0\exp(\sigma)$ which $\to\infty$ as $t\to\infty$;

(vi) see Figure 12.4 for $\widetilde{\Delta}[k,t]$ as a function of k/k_0, for three values of σ: $\sigma=3/2, 3, 6$ where the middle value yields the temporal maximum of the spectral peak;

(vii) the essential point is that the initial patchiness in the transformed concentration D at wavenumber k_0, imposed by the patchiness in the initial growth rate r_I, reaches a maximum at $t\sim\lambda^{-1/3}$ and then becomes lost as significant variance accumulates at ever increasing wavenumbers. Bennett and Denman (1985) also consider a patchy growth rate field specified independently of the velocity field, such as a nutrient introduced by local upwelling. Patchiness appears in the early transformed concentration field but is again lost as variance cascades to higher wavenumbers.

13
Diffusion

13.1 Scalar diffusion: An approximate general solution

The Eulerian conservation law for a diffusing scalar concentration $C[x_j, t]$ in a compressible flow is the well-known advection diffusion equation

$$\frac{\partial}{\partial t}C[x_j, t]+u_m[x_j, t]\frac{\partial}{\partial x_m}C[x_j, t]=S[x_j, t]+\kappa\frac{\partial^2}{\partial x_n \partial x_n}C[x_j, t], \qquad (13.1)$$

where $u_m[x_j, t]$ is as always the Eulerian fluid velocity, $S[x_j, t]$ is the source strength and κ is the molecular diffusivity (e.g., Chapman and Cowling, 1970). The derivation of the Lagrangian form of (13.1) will now be reviewed.

If a particle is launched at the point q_j at time v, then its position x_i at some other time t is a function of q_j, v and t:

$$x_i = X_i(q_j, v|t), \qquad (13.2)$$

where of course

$$\frac{\partial}{\partial t}X_i(q_j, v|t)=u_i\big[X_l(q_j, v|t), t\big] \qquad (13.3)$$

and

$$X_i(q_j, v|v)=q_i\,. \qquad (13.4)$$

Equally, if the particle is known to be at position x_i at time t, then its position q_j at some other time v is

$$q_j = X_j(x_i, t|v)\,. \qquad (13.5)$$

Using q_j and v to denote the Eulerian coordinates for space and time, respectively, the Eulerian equation (13.1) is simply changed to

$$\frac{\partial}{\partial v}C[q_j, v]+u_m[q_j, v]\frac{\partial}{\partial q_m}C[q_j, v]=S[q_j, v]+\kappa\frac{\partial^2}{\partial q_n \partial q_n}C[q_j, v]. \quad (13.6)$$

This will now be transformed into a Lagrangian equation with labeling coordinates x_i, t.

In the notation of Kraichnan (1965),

$$C(x_i, t|v) \equiv C\big[q_j, v\big]\Big|_{q_j=X_j(x_i,t|v)}, \quad (13.7)$$

and in light of (13.4),

$$C(x_i, t|t)=C[x_i, t]. \quad (13.8)$$

Exercise 13.1 Show that

$$\frac{\partial}{\partial v}C(x_i, t|v)=\kappa\left(R_l\frac{\partial}{\partial x_l}C(x_i, t|v)+Q_{lm}\frac{\partial^2}{\partial x_l \partial x_m}C(x_i, t|v)\right)+S(x_i, t|v), \quad (13.9)$$

where

$$R_l=\frac{\partial^2}{\partial q_p \partial q_p}X_l(q_j, v|t)\Bigg|_{q_j=X_j(x_i,t|v)}, \quad (13.10)$$

$$Q_{lm}=\left[\tfrac{\partial}{\partial q_p}X_l(q_j, v|t)\tfrac{\partial}{\partial q_p}X_m(q_j, v|t)\right]\Big|_{q_j=X_j(x_i,t|v)}. \quad (13.11)$$

Hint: $C[q_j, v]=C\big(X_i(q_j, v|t), t|v\big)$. □

Exercise 13.2 Show that the Jacobi matrix element appearing in (13.9)–(13.11) evolves with the flow according to

$$\frac{\partial}{\partial t}\frac{\partial X_l}{\partial q_p}=\frac{\partial u_l}{\partial x_m}\frac{\partial X_m}{\partial q_p}. \quad (13.12)$$

Note that the velocity gradient in (13.12) is Eulerian. Let δq_l be an infinitesimal line of fluid particles at time v; show that it evolves as

$$\frac{\partial}{\partial t}\delta X_l=\frac{\partial u_l}{\partial x_m}\delta X_m, \quad (13.13)$$

where $\delta X_l=\delta q_l$ at $t=v$. Thus the time-scale for infinitesimal line stretching in isotropic turbulence is $\Omega_0^{1/2}$, where Ω_0 is the rms strain rate; see (12.47).

□

The Lagrangian form (13.9) for the scalar diffusion equation is no more easily integrated than the Eulerian form (13.1). The lhs of (13.9) is the simpler, but the rhs is the more complicated owing to the spatial dependence of the transformation factors appearing in the Lagrangian form of the Laplace operator. However, if that spatial dependence is ignored, (13.9) is readily integrated. For simplicity assume vanishing concentration at $t=s$:

$$C[x_i, s]=0. \tag{13.14}$$

Exercise 13.3 Express the solution of (13.9), (13.14) using Fourier transforms. In particular, show that

$$C[x_i, t]=C(x_i, t|t)=\int_s^t \int S(y_i, t|v)G(x_i-y_i, t|v)dv dV(y), \tag{13.15}$$

where the Fourier transform of the influence function $G(x_i, t|v)$ is

$$\widetilde{G}(k_i, t|v)=\exp\left(-\int_v^t W(t|w)dw\right), \tag{13.16}$$

where

$$W(t|r)=\kappa\{ik_l R_l(\quad, t|v)+k_l Q_{lm}(\quad, t|v)k_m\}. \tag{13.17}$$

The void in the arguments $(\quad, t|v)$ implies that the spatial dependence has been ignored. □

The solution (13.15) as sketched in Figure 13.1 shows the particle path $(X_j(x_i, t|v), v)$ of a fluid particle, and the diffusion cloud $G(y_i-x_i, t|v)$. The latter indicates the advective-diffusive spread of the scalar concentration to x_i at time t, following injection at $X_i(y_j, t|v)$ at time v.

Ignoring the spatial variability of the coefficients in (13.9) is justified in three circumstances:

(i) *either* scalar diffusion is much faster than the evolution of the transformation factors R_l and Q_{lm}, that is, much faster than infinitesimal line stretching: equilibrium between the external scalar source and the diffusion sink is attained before the transformation factors have altered significantly from their spatially uniform initial values

$$R_l(x_i, t|t)=0, \qquad Q_{lm}(x_i, t|t)=\delta_{lm}; \tag{13.18}$$

(ii) *or* infinitesimal line stretching owes principally to velocity shears with scales much larger than those at which scalar diffusion is significant, thus stretching rates are approximately uniform in space;

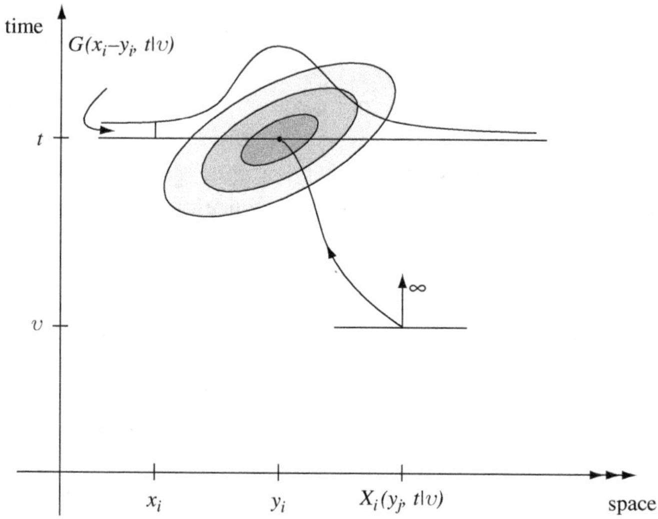

Figure 13.1 Graphical representation of the solution (13.15). The scalar injected by the source S at $X_i(y_j, t|v)$, at time v, has diffused into an ellipsoidal cloud about y_i at time t. The cloud concentration at x_i, at time t, is proportional to $G(x_i - y_i, t|v)$. The vertical arrow through $X_i(y_j, t|v)$ indicates that $G(x_i - y_i, v|v) = \Pi_{k=1}^{3}\delta(x_k - y_k)$. In this graphical representation, the diffusing cloud has been rotated 90° out of the space manifold to which it properly belongs. After Bennett (1987).

(iii) *or* only upper bounds for the scalar variance are required, and adequate bounds may be obtained without having to admit nonuniform line stretching.

It is emphasized that the particle paths along which the source S in (13.15) is being sampled will *not* necessarily be assumed to be those found in a uniform shear flow.

13.2 Variance spectrum

Assume that the velocity field u_j and source strength S are independent, isotropic and stationary random fields. Assume further that their expectation values vanish unless otherwise stated:

$$E\{u_j\} = 0, \quad E\{S\} = 0. \tag{13.19}$$

The covariance of scalar concentration at time t is

$$E\{C[x_i+\xi_i, t]C[x_i, t]\} = E_{u,S}\left\{ \int_s^t \int_s^t \iint S[X_j(y_i, t|v), v]S[X_l(z_i, t|w), w] \right.$$
$$\times G(x_i+\xi_i - y_i, t|v)G(x_i - z_i, t|w)$$
$$\left. \times dV(y)dV(z)dv dw \right\}, \tag{13.20}$$

where $E_{u,S}$ indicates that the expectation is taken over the independent ensembles of velocity and source strength. Taking the expectation over S first, and assuming a temporally white noise source as in (12.6), leaves

$$E\{C[x_i+\xi_i, t]C[x_i, t]\} = E_u\left\{ \int_s^t \iint \Sigma[X_j(y_i, t|v) - X_j(z_i, t|v)] \right.$$
$$\left. \times G(x_i+\xi_i - y_i, t|v)G(x_i - z_j, t|v)dV(y)dV(z)dv \right\}, \tag{13.21}$$

where again Σ is the spatial covariance of S. It will be convenient to replace the coordinates (y_i, z_i) at time t with their centroid and separation (c_i, f_i). Note that the argument of Σ in (13.21) depends upon the separation at time v:

$$r_j = X_j\left(c_n + \frac{1}{2}f_n, t|v\right) - X_j\left(c_n - \frac{1}{2}f_n, t|v\right), \tag{13.22}$$

but not upon the centroid at time v. Thus the expectation $E_u\{\ \}$ in (13.21) may be expressed in terms of a marginal pdf. The random variables are the separation r_i, and the transformation factors $R_l(, t|v)$, $Q_{lm}(, t|v)$ in the influence function G:

$$E\{C[x_i+\xi_i, t]C[x_i, t]\} = \int_s^t \int \cdots \int \Sigma[r_j]$$
$$\times G\left(x_i+\xi_i - c_i - \frac{1}{2}f_i, t|v\right) G\left(x_i - c_i + \frac{1}{2}f_i, t|v\right)$$
$$\times P(f_i, 0, \delta_{lm}, t|r_i, R_l, Q_{lm}, v)dV(r)dV(R)dV(Q)$$
$$\times dV(f)dV(c)dv. \tag{13.23}$$

Note:

(i) P is independent of c_i, the centroid at time t, as a consequence of assuming statistical homogeneity;
(ii) the values of R_l and Q_{lm} at time t are 0 and δ_{lm}, respectively;
(iii) the volume element $dV(Q)$ for Q_{lm} in (13.23) is in $D(D+1)/2$ dimensions, since the $D \times D$ matrix is symmetric.

Exercise 13.4 Show from (13.23) that the one-dimensional spectrum of scalar variance is

$$F[k,t] = \int_s^t \iiiint\!\!\int \Sigma[r_j] \left|\widetilde{G}(k_j, t|v)\right|^2 e^{ik_j f_j} \times P(f_j, 0, \delta_{lm}, t|r_j, R_l, Q_{lm}, v)\, dV(f)\, dV(r)\, dV(R)\, dV(Q)\, dA(k)\, dv\,, \tag{13.24}$$

where again

$$F[k,t] \equiv \iint E\left\{C[x_j+\xi_j, t]C[x_j, t]\right\} e^{ik_j\xi_j}\, dV(\xi)\, dA(k)\,, \tag{13.25}$$

and $\widetilde{G}$ is the Fourier transform of G. If the scalar diffusivity κ is set to zero then $G \equiv 1$. In that case, after integrating over the marginal random variables R_l and Q_{lm}, and after exploiting isotropy, (13.25) becomes

$$F[k,t] = A(k) \int_s^t \int_0^\infty \int_0^\infty \mathcal{S}(kf) A(f) A(r) \Sigma[r] P(f, t|r, v)\, dr\, df\, dv\,, \tag{13.26}$$

which is just (12.13), (12.16) again. □

13.3 Enstrophy inertia diffusive subrange

Consider the enstrophy-cascading inertial subrange of two-dimensional turbulence. It will be seen in subsequent sections that the rate of infinitesimal line stretching is characterized by the root-mean-square strain rate, that is, by $O(\lambda^{1/3})$ here,[1] where λ is the enstrophy cascade rate. The logarithmic separation rate for particle pairs, defined as $r^{-2}\eta$, is also $O(\lambda^{1/3})$ according to (12.27). However, infinitesimal line stretching is controlled by the rms strain field (summed over all wavenumbers) while separation in the inertial subrange is influenced by the local wavenumber k, so the two processes

[1] with logarithmic corrections.

are statistically indendent. The rate of infinitesimal line stretching is greatly exceeded by the scalar diffusion rate κk^2 if

$$\lambda^{1/6}\kappa^{-1/2} \ll k\,. \tag{13.27}$$

On the other hand, this subrange is bounded above by the upper limit of the enstrophy-cascading inertial subrange of the turbulence. Let us suppose that (13.27) does hold. Then infinitesimal line stretching is negligible, and so the transformed influence function may be approximated:

$$\widetilde{G}(k_j, t|s) \cong \exp\left[-\kappa k^2(t-s)\right]. \tag{13.28}$$

In particular, the fundamental assumption in (13.16) of spatially uniform stretching factors is trivially justified, as these factors do not evolve significantly from their uniform initial values (13.18) during the diffusion process. Only the short-time statistics of separation are required, and a short-time ($t \approx s$) approximation to the longitudinal diffusivity suffices. Indeed, as a first approximation it would seem appropriate to approximate the separation pdf $P(f, t|r, s)$ by its initial form $A(f)^{-1}\delta(f-r)$. However, this leads immediately to the scalar variance spectrum

$$F[k, t] \sim A(k)\widetilde{\Sigma}[k](\kappa k^2)^{-1}\,, \tag{13.29}$$

as $t-s \to \infty$, where $\widetilde{\Sigma}[k]$ is the Fourier transform of the source covariance. That is, owing to the neglect of relative dispersion, there is no cascade of scalar variance away from the source wavenumbers. It is evidently necessary to admit that P has a small but finite spread about $r = f$. The spread variance is $O\left(\lambda^{2/3}(t-s)^2 f^2\right)$, and the resulting correction to (13.29) depends upon the detailed form of $\widetilde{\Sigma}$; that is, F has no *universal* form.

The preceding analysis has assumed that the scalar is sustained by an isotropic source. This is an idealization; a more realistic (and more easily realized) model has no external source of scalar variance, but is instead sustained by a mean scalar concentration with a gradient that is uniform in space and in time. Without loss of generality it may be assumed that the gradient is parallel to one of the space axes:

$$\nabla E\{C\} = (\Gamma, 0, 0)\,. \tag{13.30}$$

In this model the turbulence is still assumed to be stationary, isotropic and having zero mean. Fluctuations in C can be induced by turbulent advection of the mean scalar gradient or by random initial values of C, but the latter possibility will be ignored by assuming $C' \equiv C - E\{C\} = 0$ at $t = s$.

It is easily seen that C' satisfies the advection diffusion equation (13.1), with the random source S replaced by $-u_1[x_j, t]\Gamma$. Hence the solution, according to the representation (13.15), is

$$C'[x_i, t] = -\Gamma \int_s^t \int G(x_i - y_i, t|v) u_1(y_i, t|v) dv dV(y) . \tag{13.31}$$

The inertia diffusive approximation (13.28) will again be assumed. An approximate expression for the scalar variance spectrum $F[k, t]$ may now be developed. The expression includes the Lagrangian velocity correlation $E\{u_j(y_l, t|v)u_j(z_m, t|w)\}$ for $s < v < t$ and $s < w < t$. However, it is only necessary to consider $t - w$ and $t - v$ both $O(\kappa k^2)^{-1}$, which time-scale is much shorter than the $O(\lambda^{-1/3})$ decorrelation time of the velocity field. Thus the Lagrangian covariance may be approximated by the Eulerian covariance $E\{u_i(y_l, t|t)u_j(z_m, t|t)\} = E\{u_j[y_l, t]u_j[z_m, t]\}$. Hence, asymptotically for large t, and assuming stationary isotropic turbulence, the approximate scalar spectrum is

$$F[k] \sim \Gamma^2 E[k](\kappa k^2)^{-2} . \tag{13.32}$$

The energy spectrum $E[k]$ for the enstrophy subrange is given by (11.47), hence

$$F[k] = K_r \Gamma^2 \lambda^{2/3} \kappa^{-2} k^{-7} . \tag{13.33}$$

No observations are available for testing (13.33). These would have to be collected in a laboratory on the small scale, rather than in the ocean or atmosphere on the large scale, so that the diffusion of the scalar is indeed molecular as in (13.1).

13.4 Energy inertia diffusive subrange

Infinitesimal line stretching in isotropic turbulence proceeds at the rms strain rate $\Omega_0^{1/2}$, which is $O(\epsilon/\nu)^{1/2}$ for an energy cascade. Within the energy inertial subrange itself, particle pairs with separation $r \sim k^{-1}$ separate at the rate $\epsilon^{1/3}k^{2/3}$. Let $k_\kappa = \epsilon^{1/4}\kappa^{-3/4} = k_\nu P_r^{3/4}$. If $P_r \ll 1$, then there is an inertia diffusive subrange, $k_\kappa \ll k \ll k_\nu$, between the inertia convective and viscous diffusive subranges. In this intermediate subrange, $\nu k^2 \ll \epsilon^{1/3}k^{2/3} \ll \kappa k^2$. That is, the diffusion rate exceeds the pair separation rate, which in turn exceeds the viscous dissipation rate. Pair separation is slower than infinitesimal line stretching $(\epsilon^{1/3}k^{2/3}/\Omega_0^{1/2} \sim (k/k_\nu)^{2/3} \ll 1)$; more importantly, the former is local in wavenumber space at wavenumber $k \sim r^{-1}$ while the latter is local in real space and so, as in the enstrophy inertia diffusive subrange, the two processes are statistically independent. If this interval is restricted to $k_B = k_\nu Pr^{1/2} \ll k \ll k_\nu$, then $\Omega_0^{1/2} \ll \kappa k^2$. That is, the rate of infinitesimal line stretching is much less than the scalar diffusion rate and so the transformed influence function once more has the simple form

$$\widetilde{G}(k_j, t|s) \cong \exp\left[-\kappa k^2(t-s)\right]; \tag{13.34}$$

again the statistics of stretching are not required.

The isotropic source model does not lead to a universal form for the scalar spectrum. The uniform gradient model leading to (13.32), combined with the energy inertial subrange (11.41), yields

$$F[k] \sim K_o \Gamma^2 \epsilon^{2/3} \kappa^{-2} k^{-17/3}. \tag{13.35}$$

This result owes originally to Batchelor et al. (1959); it is also derived by Kraichnan (1968) using his Lagrangian History Direct Interaction Approximation. More recently, Lesieur et al. (1981) and Lesieur and Herring (1985) derive (13.33) and (13.35) using an "eddy damped quasi-normal" closure theory.

The just-mentioned other derivations assume an isotropic source, but in effect argue that a low-wavenumber component of the scalar field is in

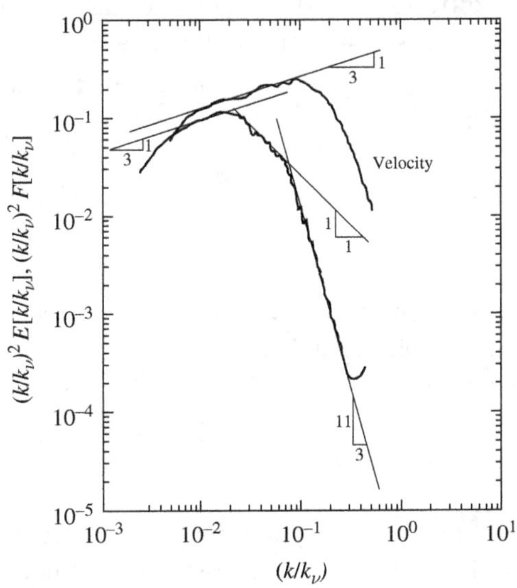

Figure 13.2 Scaled variance spectra $(k/k_\nu)^2 E$ and $(k/k_\nu)^2 F$, respectively, for velocity and scalar concentration in mercury *versus* k/k_ν at low Prandtl number ($Pr = 0.018$); after Clay (1973).

practice indistinguishable from a mean gradient in the field. The squared mean gradient Γ^2 is effectively identified with $\chi\kappa^{-1}$, the scaled variance of source strength. Also, the variance production $E\{CS\}$ is omitted, so is evidently a cause of nonuniversality. The observations of Clay (1973) in a laboratory channel using mercury ($P_r = 0.02$) clearly support (13.35); see Figure 13.2. Indeed, Clay's experimental configuration is better described by the mean gradient model than the isotropic source mode.

13.5 Viscous diffusive subrange

For all Prandtl numbers $P_r \equiv \nu/\kappa$, there is a viscous diffusive subrange $\max[k_\nu, k_B = k_\nu Pr^{1/2}] \ll k$, in which viscous dissipation of kinetic energy and diffusion of scalar concentration variance are both significant. Infinitesimal line stretching proceeds as always at the rms strain rate (given by $\Omega_0^{1/2} \sim (\epsilon/\nu)^{1/2}$ for an energy cascade) but, as $k_\nu \ll k$, so does the separation of particle pairs. That is, both processes are local in real space. Indeed, they owe to one and the same realization of the local velocity shear, and so must be assumed statistically dependent. Thus joint statistics of stretching and separation are needed in order to estimate the scalar spectrum via (13.24), even though scalar diffusion is faster than stretching. Specifically, there is a requirement for the joint statistics of the finite separation r_j, and the transformed influence function $\widetilde{G}$. The equation (13.24) for the scalar spectrum may be reorganized as

$$F[k, t] = \int_s^t \iiint \widetilde{\Sigma}[l_j] E\left\{ |\widetilde{G}|^2 \exp\left(i(k_j f_j - r_j l_j)\right)\right\} dV(l) dV(f) dA(k) dv, \tag{13.36}$$

where $\widetilde{\Sigma}$ is the Fourier transform of Σ. The expectation is taken with respect to the joint random variables r_j and $\widetilde{G}$.

The separation r_j satisfies

$$\frac{\partial}{\partial v} r_j = u_j[q_i + r_i, v] - u_j[q_i, v] \tag{13.37}$$

subject to $r_j = f_j$ at $v = t$. For $r \ll k_\nu^{-1}$, separation is controlled in the rms sense by eddies at the peak of the rate-of-strain spectrum $k^2 E[k]$, that is of scale $k^{-1} = k_\nu^{-1} \gg r$. Hence r_j obeys, essentially,

$$\frac{\partial}{\partial s} r_j = Z_{jm} r_m, \tag{13.38}$$

where

$$Z_{jm} = \left[\frac{\partial}{\partial q_m} u_j[q_n, v] \right] \Bigg|_{q_n = X_n(x_i, t|v)}, \tag{13.39}$$

and $r_j = f_j$ at $v = t$. The separation rate for pairs in this wavenumber range of homogeneous turbulence is clearly the rms strain rate: $\Omega_0^{1/2} \sim (\epsilon/\nu)^{1/2}$ for an energy cascade.

The formal solution of (13.38) is

$$r_j(t|v) = H_{jm}(t|v) f_m, \tag{13.40}$$

where the propagator H_{jm} satisfies

$$\frac{\partial}{\partial s} H_{jm}(t|v) = Z_{jn}(t|v) H_{nm}(t|v), \tag{13.41}$$

subject to $H_{jm} = \delta_{jm}$ at $v = t$.

Note that the spatial arguments of Z_{jn}, and hence those of r_j and H_{jm}, have been suppressed in accordance with the representation (13.15)–(13.17). Substituting (13.40) into the exponent in (13.36) and integrating over f_j leads to

$$F[k, t] = \int_s^t \iiint \widetilde{\Sigma}[l_j] E\left\{ |\widetilde{G}|^2 \delta(k_m - H_{nm}(t|v) l_n) \right\} dV(l) dA(k) dv. \tag{13.42}$$

So the expectation in (13.42) is the variance of the amplitude of $\widetilde{G}$, conditional upon the wavenumber k_m taking time-dependent values. It is therefore convenient to introduce a function of time in wavenumber space. In the interest in minimizing the number of symbols, the function will be denoted $k_n(t|s)$, where

$$\frac{\partial}{\partial t} k_n(t|s) = -Z_{pn}(s|t) k_n(t|s). \tag{13.43}$$

Exercise 13.5 Show that H_{jm}, the propagator for evolution of r_j in s, satisfies

$$H_{jn}(t|s) H_{nm}(s|t) = \delta_{jm}, \tag{13.44}$$

and hence is also the propagator for k_n in t:

$$k_n(t|s) = H_{jn}(t|s) l_j, \tag{13.45}$$

where $l_j = k_j(t|t)$. □

Next, an evolution equation will be derived for the influence function $\widetilde{G}$. It follows from (13.16)–(13.18) that

$$\frac{\partial}{\partial t}\widetilde{G}(k_j, t|s) \cong -\kappa k^2(t|s)\widetilde{G}(k_j, t|s)\,. \tag{13.46}$$

Recall that in the viscous diffusive subrange ($k_B = k_\nu Pr^{1/2} \ll k$), the stretching rate is much smaller than the rate of scalar diffusion: $\Omega_0^{1/2} \ll \kappa k^2$. Now according to (13.46), $\widetilde{G}$ is real-valued since $\widetilde{G}=1$ at $t=s$. Hence the expectation in (13.42) is $E\{\widetilde{G}^2\}$ where $\widetilde{G}$ and k_n together satisfy the system (13.43), (13.46), subject again to $(\widetilde{G}, k_n) = (1, l_n)$ at $t=s$.

Before proceeding, it is appropriate to review what has transpired in the derivation of (13.46). It would seem that no evolution of the transformation factors R_l and Q_{lm} in (13.17) has been allowed, since they have been kept at their initial values of 0 and δ_{lm}, respectively. However, the wavenumber k_j in (13.17) now evolves according to (13.43), which evolution it has inherited from the separation propagator H_{lm} defined in (13.38). Thus the question becomes: Is this propagator appropriate also for the transformation factors? Diffusion is so fast in this subrange that only the evolution of Q_{lm} need be considered, since R_l vanishes initially.

Exercise 13.6 Define the wavenumber function h_n by

$$h_n = l_j\left(\frac{\partial}{\partial q_n}X_j(q_m, s|t)\right)\Bigg|_{q_m = X_m(x_i, t|s)}, \tag{13.47}$$

where l_j is a fixed wavenumber. Hence $h_n = l_n$ at $t=s$. It is clear from (13.11) that

$$h^2 = h_n h_n = l_j Q_{jp} l_p\,. \tag{13.48}$$

Then use the labeling theorem to show that

$$\frac{\partial}{\partial s}h_n = -Z_{np}(t|s)h_p\,. \tag{13.49}$$

□

Comparing (13.43) and (13.49), it may be concluded that the evolution of the statistics of the transformation factor Q_{jp} is faithfully retained, so long as $Z_{pn}(s|t)$ and $Z_{np}(t|s)$ are statistically indistinguishable.

Kraichnan (1974) arrives at (13.43), (13.46) with an analogous construction. He then proposes a one-dimensional stochastic model for $k = \sqrt{k_n k_n}$:

$$\frac{dk}{dt} = \alpha k\,, \tag{13.50}$$

with $k=l$ at $t=s$. The white-noise strain rate $\alpha(t)$ has positive mean μ and variance σ^2. That is, (13.50) should be expressed as the stochastic differential equation.

$$dk=\mu k dt+\sqrt{2}\sigma k d\beta, \tag{13.51}$$

where $d\beta(t)$ is the Wiener process of unit variance (e.g., Gardner, 1985; van Kampen, 1992; Rodean, 1996). Both μ and σ^2 are $O(\Omega_0^{1/2})$. Kraichnan argues that $\mu=D\sigma^2$, where D is the number of space dimensions in (13.43). Once the joint pdf $P(1, l, s|\widetilde{G}, k, t)$ for the system (13.46), (13.51) has been found, and assuming for simplicity that $\widetilde{\Sigma}[l_j]=\chi\delta(l_j-m_j)$, the equilibrium scalar spectrum is

$$F[k, t]\sim\chi\int_0^\infty\int_0^1 P(1, m, 0|g, k, t)g^2 dg dt, \tag{13.52}$$

where $m=\sqrt{m_j m_j}$. The notation $g\equiv\widetilde{G}$ is introduced in (13.52) for subsequent convenience. Note that g takes all values in the interval $[0, 1)$. The joint pdf in (13.52) satisfies the Fokker–Planck equation

$$\frac{\partial}{\partial t}P=-\mu\frac{\partial}{\partial k}(kP)+\kappa k^2\frac{\partial}{\partial g}(gP)+\sigma^2 k\frac{\partial}{\partial k}\left(k\frac{\partial}{\partial k}P\right), \tag{13.53}$$

according to the Stratonovitch interpretation, in which $\alpha(t)$ is regarded as a process with a positive but vanishingly small decorrelation time (Rodean, 1996, etc.). The initial condition is

$$P(1, m, 0|g, k, 0)=\delta(g-1)\delta(k-m). \tag{13.54}$$

Steady state solutions of (13.53) have been obtained by Kraichnan (1974), time-dependent solutions subject to (13.54) by Bennett (1986a). These solutions are particularly simple in the case $\sigma=0$, which corresponds to Batchelor's (1959) uniform strain model. The equilibrium spectrum is then

$$F[k, t]\sim\chi\mu^{-1}k^{-1}\exp\left(-\kappa\mu^{-1}(k^2-m^2)\right), \tag{13.55}$$

as given by Batchelor. If $\sigma>0$, then the equilibrium spectrum is of the form

$$F[k, t]\sim F[k]\propto\chi k^w\exp\left(-(2\kappa\sigma^{-1})^{1/2}k\right), \tag{13.56}$$

as $k\to\infty$, $m\to 0$ (Kraichnan, 1974), where $w=(\mu\sigma^{-1}-3)/2$. However, if the range of integration over g in (13.52) is restricted to $g_0\le g\le 1$, then, still for the case $\sigma>0$, the equilibrium spectrum is of the form

$$F[k]\propto\chi k^{2w}(\ln g_0)^{w-1/2}\exp\left(-\kappa\mu^{-1}(4\sigma\ln g_0)^{-1}k^2\right). \tag{13.57}$$

This is essentially the same form as the Batchelor spectrum (13.55) even though the strain rate α is now random. The cutoff g_0 may represent the threshold sensitivity of measurement.

The adoption of a white-noise model for the strain rate $\alpha(t)$ as in (13.50), (13.51) is extreme. Consider instead the model

$$\frac{dk}{dt}=k\Omega_0^{1/2}M(\theta)\,, \tag{13.58}$$

where θ is a standard normal random variable and M is some positive-valued functional form. It is assumed that θ is independent of time, or else is a stationary process with a very long decorrelation time. The equilibrium spectrum $F_\theta[k]$ for a particular realization of θ is given by the Batchelor form (13.55) with μ replaced by $\Omega_0^{1/2}M(\theta)$. Then $F[k]$ is obtained by averaging over θ. Asymptotic forms may be obtained using the method of steepest descent, as $k\to\infty$. There are several interesting examples.

(a)

$$M(\theta)=\exp(\theta)\,. \tag{13.59}$$

This log normal strain rate is a natural choice. The spectrum is

$$F[k]\propto\chi\Omega_0^{-1/2}k^{-2}\ln(k/k_B)\exp\left(-\frac{1}{2}\left(\ln(k/k_B)\right)^2\right)\,. \tag{13.60}$$

Note that $k_B^2=\Omega_0^{1/2}\kappa^{-1}$. Thus, intermittency of the strain rate leads to a very broad spectrum.

(b)

$$M(\theta)=O(1)\,,\ \frac{d}{d\theta}M(\theta)=O(1)\ \text{as}\ |\theta|\to\infty\,. \tag{13.61}$$

Then for large k,

$$F[k]\propto\chi\Omega_0^{-1/2}k^{-1}\exp\left(-b_1(k/k_B)^2-b_2(k/k_b)^4\right)\,, \tag{13.62}$$

where b_1, b_2 are positive, bounded dimensionless functions of k/k_B. This resembles the Batchelor form (13.55).

The parade of scalar spectra: (13.52), (13.53), (13.54), (13.57) and (13.59) may be compared with observations. Gargett (1985) provides a review, and presents high-quality data from a turbulent coastal channel; see Figure 13.3. These newer observations support the Batchelor spectrum (13.55), although no universal value is found for $\Psi\equiv\Omega_0^{1/2}\mu^{-1}$. Note that the exponential in (13.55) is $\exp\left(-\Psi(k^2-m^2)k_B^{-2}\right)$. For large signal-to-noise ratios (small g_0), Gargett finds large values for Ψ, which contradicts (13.57). The most

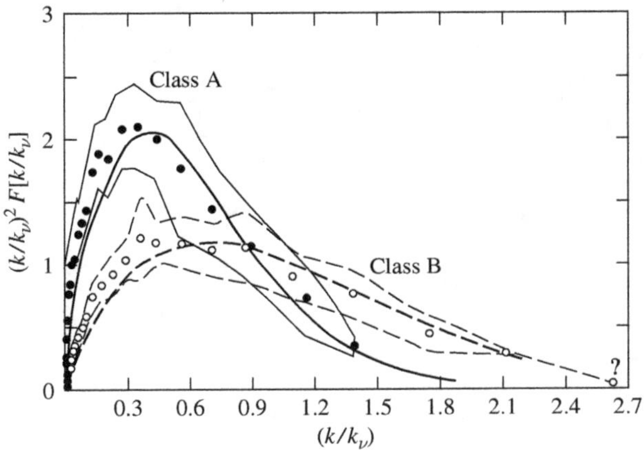

Figure 13.3 Variance-preserving plot of temperature variance dissipation spectra in a British Columbia fjord. *Heavy solid curve*: Batchelor spectrum (13.55) with $\Psi \equiv \Omega_0^{1/2}\mu^{-1} = 12$. *Heavy dashed curve*: $\Psi = 4$. Note that, with reference to the exponent in (13.55), the coefficient is $\kappa\mu^{-1} = \Psi/k_B^2$. After Gargett (1985).

plausible model is (13.62); if the true spectrum were of this form, then fitting the Batchelor form to the data (i.e., assuming $b_2 = 0$) would lead to an overestimate for b_1. The safest conclusion is that while theoretical models of the viscous diffusive subrange are highly sensitive to model details, Batchelor-like spectral forms are ubiquitous but universality is not likely.

PART IV

Lagrangian Data

Introduction

A benefit of the great majority of Lagrangian data being so new is that most of them are available via the Internet. Also, Lagrangian time series analysis may be carried out using conventional time series methods, for which well-supported software libraries abound. Modern dynamical systems theory suggests new and intriguing quantities characterizing the behavior of Lagrangian time series.

Combining Lagrangian data with Lagrangian dynamical models can be conceptually as simple as combining Eulerian data with Eulerian dynamical models, and as complex to implement effectively. Mixing the two formulations of fluid dynamics leads to nonlinearities in the measurement functionals, further complicating effective implementation. Particle pairs make incisive tools for investigating the field of flow: dynamically constrained analysis of float pairs, without having to run all the machinery that is a modern ocean circulation model, is theoretically possible and offers the cability for real-time, even onboard analysis.

14
Observing systems

14.1 The laboratory

Measurements of the displacements of fluid particles provide locally averaged values of Lagrangian velocity. With first-order accuracy, these averages are also the local Eulerian velocity:

$$X_i(a_j, s|s+\delta t) - X_i(a_j, s|s) = \int_s^{s+\delta t} u_i(a_j, s|r)dr$$
$$= \delta t u_i[a_j, s] + \mathrm{O}(\delta t)^2 . \qquad (14.1)$$

The truncation error can be made very small by reducing the time interval, but then the experimental accuracy of the measurement of displacements may become as large. In any event, these are only local measurements. Particles are almost never tracked in the laboratory for significant times, owing to the difficulty in following an individual particle. Nevertheless, almost-Lagrangian velocimetry is both powerful and fascinating. Ingenious techniques have been devised for marking fluid particles, such as pulsing current through wires in the fluid which is actually an aqueous solution of a pH indicator buffered to the neutral point (Baker, 1966). For applications, see for example Beardsley (1969), Baker and Robinson (1969). More recent marking techiques are mentioned in Guyon, Hulin, Petit and Mitescu (2001). A directory of commercial Web sites has been compiled by David Holland (personal communication, 2003).[1] Software for particle-imaging velocimetry has been developed at the Coriolis facility in Grenoble, and is freely available.[2] The Grenoble system is designed for massive image processing, as is needed for turbulence studies of

[1] `http://fish.cims.nyu.edu/laboratory/lab_physical/lab_suppliers.html`

[2] `http://www.civproject.org`

three-dimensional velocity measurements throughout a volume (J. Sommeria, personal communication, 2003).

14.2 The atmosphere

Wind, temperature and dew point are routinely measured from expendable weather balloons. The wet bulb and dry bulb temperature data are transmitted to the ground, by an expendable radio, at regular intervals during the ascent of the balloon. Assuming a uniform rate of ascent yields moist thermodynamic profiles of the lower atmosphere up a height of about 3 km. Wind is inferred by tracking the balloon with theodolites[3] or radar. Range and accuracy requirements vary according to application, which may include fighting forest fires as well as preparing synoptic-scale meteorological analyses. Interesting Web sites abound.[4]

The TWERL experiment conducted during the First GARP Global Experiment, which data are discussed in Chapter 11, involved pairs of high-altitude balloons tracked by earth satellites (Julian, Massman and Levanon, 1977). These remarkable observations, which have yielded so many tantalizing insights in geophysical turbulence, have never been repeated.[5]

14.3 The ocean surface

The measurement of surface currents has been intrinsically Lagrangian since the earliest times. For an engaging review, see Gould (2001). However, the determination of the drift track of a drogued buoy was, until the advent of earth satellites, dependent upon knowing the observing ship's position and was subject to many uncertainties (Tizard, Moseley, Buchanan and Murray, 1885):

> These results were assumed as giving the rate and direction of the currents at different depths with sufficient accuracy to ascertain any marked movements, but it is evident that they are not strictly accurate, as no allowance was made for the retarding or accelerating influence of the surface water on the watch buoy, or of the intermediate water on the line.

[3] http://www.warrenind.com/WKMeteorology.html#Pilot%20balloon%20Theodolite
[4] http://www.srh.noaa.gov/elp/kids/balloon.html
[5] http://dss.ucar.edu/download/nmc/twerle

Over a century later Gould (1998) is reporting that

> Surface drifters were already used in large numbers when WOCE (the World Ocean Circulation Experiment) was planned. However their ability to represent upper ocean currents unambiguously and to survive for multi-year missions was poor. Developments to meet TOGA (the Tropical Ocean-Global Atmosphere research program) and WOCE requirements have resulted in lower cost and improved data quality and quantity.

Advances in drifting buoy technology are described by Kennan, Niiler and Sybrandys (1998). They remark that

> Since winds cause drifters to slip through the water, it is desirable to have subsurface drogues to follow the motion representative of near surface circulation. However, a surface float, which is inevitably subject to the extremes of winds stress, seas, and swell, is required so that the drifter can telemeter its observations. Furthermore, vertical shear of the near surface currents and wave forces induce variable slip along the length of the drogue (Niiler, Davis and White, 1987). Consequently, upper ocean drifters are not perfect Lagrangian parcels.

They go on to describe the SVP (Surface Velocity Program) drifter design for WOCE, TOGA and CLIVAR (the CLImate VARiability research program)[6], and they also develop a simple mathematical model for drifter slip velocity U_S expressed as a complex number:

$$U_S = \left(ae^{i\alpha}W + be^{i\beta}\delta U\right)R^{-1}, \tag{14.2}$$

where W is the complex wind velocity at height of 10 m, δU is the complex velocity shear of horizontal currents separated by the length of the drogue, while the real number R is the ratio of the drogue drag area to the sum of the drag areas of the float and tether (drag area equals drag coefficient multiplied by area). The angles of slip relative to the wind and shear are α and β, respectively. Niiler et al. (1995) report that

> over 84% of the variance in the slip ... can be accounted for by linear fits to the four coefficients (a, b, α, β), giving the result that that R must be greater than 40 to achieve less than 1 cm/s slip in 10 m/s winds (performance in stronger winds is unknown).

Later, Niiler et al. (1995) remark

> it follows that knowledge of the winds can be used to correct drifter motions for slip ... Drifter deployment may take place from ships or even aircraft; once in the water, the drifter package dissolves and the drogue unfolds itself under the

[6] `http://www.aoml.noaa.gov/phod/dac/gdc.html`

> influence of gravity. The drifters telemeter their identifiers and measured parameters to polar orbiting satellites from which Service Argos produces a raw data set of buoy fixes. These data are then routinely processed, archived,and distributed by the Global Drifter Data Center at the NOAA Atlantic Oceanographic and Meteorological Laboratory (AOML), which also aids in the global deployment of drifters. Over 17 countries and 41 principal investigators have contributed data and resources to the Global Drifter Program.

The location accuracy of global Lagrangian drifters is presently estimated by AOML to be about 500 m (M. Bushnell, unpublished manuscript).[7] For a careful review of the performance of drifters, see Emery and Thomson (1997, Section 1.8) A very useful drifter bibliography has been compiled by P. Niiler and C. Martin (unpublished manuscript), most recently updated in October, 2004.[8] The Global Drifter Program has been recently summarized by Niiler (2001).

14.4 The deep ocean

The history of deep floats is also engagingly told by Gould (2001).[9] John Swallow, with his experience in developing marine seismic techniques, pioneered tracking floats acoustically. He also put into practice

> the underlying principle ... that a pressure vessel ... less compressible than sea water could be ballasted to sink at the ocean surface and would gain buoyancy relative to the surrounding water as it sank. If ballasted correctly it would find its neutral density level and drift with the currents.

Initial trials were held over the Iberian Abyssal Plain (Swallow, 1955); these were followed with neutrally bouyant float observations of the Mediterranean outflow at Gibralter (Swallow, 1969), and in the flow of Arctic water past the Faroes (Crease, 1965). The deep southward flow beneath the Gulf Stream, predicted by Stommel, was observed off North Carolina using Swallow floats (Swallow and Worthington, 1961). Swallow-float measurements of deep currents in the ocean interior revealed eddies with flow speeds of centimeters per second, and time and space scales of tens of days and tens of kilometers. Gould (2001) remarks "This was arguably the most significant discovery about the nature of the oceans in the twentieth century."

[7] `http://www.aoml.noaa.gov/phod/dac/loc-rep.html`
[8] `http://www.aoml.noaa.gov/phod/dac/drifter_bibliography.html`
[9] see also `http://www.soc.soton.ac.uk/JRD/HYDRO/argo/history.php`

Further developments in tracking and ballasting led to the RAFOS float (Rossby, Dorson and Fontaine, 1986; Rossby, Levine and Connors, 1985). This float passively tracks its position with respect to moored sound sources, and relays these data via satellite after surfacing at the end of its mission. A compressee in the float enables it to follow isopycnal surfaces. Further developments led to the ALFOS float (Ollitrault, 1993) which surfaces regularly to relay data via satellite. The WOCE Subsurface Float Data Assembly Center at the Woods Hole Oceanographic Institution[10] is a repository of float data, plus the metadata describing the floats as used and the observing programme in which they were involved. Compressee technology has advanced to the extent that it is now possible to make accurate measurement of vertical velocity in internal waves and in deep convection (D'Asaro et al., 1996; D'Asaro, 2003).[11]

The Autonomous LAgrangian Circulation Explorer or ALACE float (Davis, Webb, Regier and Dufour, 1992) dispenses with acoustic ranging. Its position is determined by satellite navigation each time that it cycles from its neutrally bouyant depth to the surface for data transmission. Thus it provides Lagrangian velocity, integrated over the time between consecutive trips to the surface. This time may vary from a few days to a month. Davis and Zenk (2001) discuss conditions in which the averaged velocity is in fact an accurate estimate of the deep circulation. The Profiling ALACE float, or PALACE float[12], samples temperature and salinity as it rises to the surface. The most recent version is the APEX float.[13]

The ideal float would be both autonomous and yet aware of its instantaneous position. Inertial guidance systems are now sufficently compact[14]; being accelerometers, they would enable second-order dead reckoning. Correlation sonar measures speed over a sufficiently rough bottom; combined with a compass, this advanced acoustic technology would provide first-order dead reckoning (D. Farmer, personal communication).

[10] http://wfdac.whoi.edu

[11] http://opd.apl.washington.edu/ dasaro/FLOTATECH/floats.html

[12] http://www.argo.ucsd.edu

[13] http://www.webbresearch.com/apex.html

[14] http//www.usna.edu/AUVT/IMU.htm

15
Data analysis: the single particle

15.1 Time series analysis: the single particle

Confronted with data exhibiting great variability, it is customary to seek to reduce the detail by regarding the data as samples from a random population or ensemble. The probability distribution function for the ensemble is assumed to be characterized by a few low-order moments which, it is hoped, may be reliably estimated from the data set. The data are likely to have been collected at a variety of times or places, and it is assumed that the more obvious characteristics of the data (amplitude range and frequency range, defined however loosely) are uniform with respect to either time or at least one spatial coordinate. Then of course sample means along the coordinate of statistical stationarity may be regarded as approximations to ensemble means. The path of a drifter on the ocean surface provides a natural first example for the estimation of ensemble parameters. The path invariably contains loops of roughly the same diameter and period of revolution. These loops are usually executed in the same sense within any one subregion of an ocean basin. The drifter velocities clearly precess in time; both the period and sense of precession are roughly uniform along the drifter path. These detailed data may be characterized by *polarization analysis*.

15.1.1 Polarization of Lagrangian velocities

Let the horizontal Lagrangian velocity compenents be

$$u = U\cos(\sigma t + \mu), \quad v = V\cos(\sigma t + \nu), \tag{15.1}$$

where U and V are constant amplitudes, σ is a real positive frequency, while μ and ν are constant phases. The labels for this single drifter have been suppressed. The manipulations which follow are expedited by introducing a complex velocity $w=u+iv$. Note that the sign convention differs from the usage in Section 6.1 for irrotational flow.

Exercise 15.1

(i) Express w in the form

$$w=a_+e^{i\sigma t}+a_-e^{-i\sigma t}. \tag{15.2}$$

(ii) Show that $|a_+|=|a_-|$ if and only if $\mu=\nu\pm n\pi$, $n=1,2,\ldots$, in which case the drifter velocity does not precess but is *rectilinearly polarized.* Show that if $a_-=0$ $(a_+=0)$, then the polarization is *circular*, that is, the drifter speed is constant, with the tip of the velocity vector describing a circle in the anticlockwise (clockwise) sense. Show that the intermediate cases amount to *elliptical* polarization.

(iii) Define the time average $T\{\ \}$, for any function of time $q(t)$, by

$$T\{q(t)\}\equiv\lim_{t\to\infty}\frac{1}{t}\int_0^t q(r)dr. \tag{15.3}$$

Evaluate $T\{w(t)\overline{w}(t+\tau)\}$ for w given by (15.2), where the overbar denotes complex conjugation, and hence find the four covariances $C_{uu}(\tau)$, $C_{uv}(\tau)$, $C_{vu}(\tau)$, $C_{vv}(\tau)$ where, for example,

$$C_{uu}(\tau)\equiv T\{u(t)u(t+\tau)\}, \tag{15.4}$$

assuming that $T\{u\}=0$.

(iv) Verify that that the skew element

$$C^A(\tau)\equiv\frac{1}{2}\Big(C_{uv}(\tau)-C_{vu}(\tau)\Big) \tag{15.5}$$

vanishes for all τ if and only if the polarization is rectilinear. Thus $C^A(\tau)$ is a measure of the polarization of the drifter velocity. Notice that $C^A(0)=0$.

(v) Show that the rate of rise of the skew element at zero lag determines the sense of polarization, with

$$\left.\frac{\partial}{\partial\tau}C^A(\tau)\right|_{\tau=0}>0 \tag{15.6}$$

for anticlockwise polarization. What are the conditions for circular polarization?

□

Exercise 15.2 Relate the polarization C^A to ϕ^A_{kl}, the antisymmetric part of the mixed diffusivity tensor appearing in the semi-empirical equation (10.81) for turbulent diffusion of the mean concentration of a passive scalar. Hence (Middleton and Loder, 1989) derive the impact of the polarization on the time rate of change of spatial moments of concentration, such as

$$S_C\{x_i x_j\} \equiv \frac{\int_{D^t} x_i x_j E\{C[x_k, t]\} dD^t}{\int_{D^t} E\{C[x_k, t]\} dD^t} . \tag{15.7}$$

Can the impact on first moments be determined from the polarization of one drifter track? □

It is clear that the time average in (15.4), for example, is a phase average of lagged products of the trigonometric functions in (15.1). Indeed, if the phases $\epsilon_\pm$ of the complex amplitudes $a_\pm$ are random, and if the phase difference $\epsilon_+ - \epsilon_-$ is uniformly distributed in $[0, 2\pi]$, then

$$E\{w(t)\overline{w}(t+\tau)\} = T\{w(t)\overline{w}(t+\tau)\} . \tag{15.8}$$

It is therefore reasonable to estimate ensemble averages by taking arithmetic means over a large number of drifters in a region that appears to be statistically homogeneous in some rough sense. With simple area weighting or volume weighting, the arithmetic means may be treated as normalized spatial integrals defined by

$$S_L\{q(t\} \equiv \frac{1}{D^s} \int_{D^s} q(a_j, s|t) dD^s , \tag{15.9}$$

where D^s is an area or volume of fluid at the labeling time s. As in Chapter 3, this Lagrangian integral may be transformed to Eulerian variables:

$$S_L\{q(t)\} = \frac{1}{D^s} \int_{D^t} q[x_i, t] J^s_t dD^t \tag{15.10}$$

where, as always, $J^s_t J^t_s = 1$. The presence of the inverse of the Jacobi determinant in (15.10) implies that the Lagrangian spatial mean $S_L\{q(t)\}$ differs from the Eulerian spatial mean $S_E\{q(t)\}$:

$$S_E\{q(t)\} \equiv \frac{1}{D^t} \int_{D^t} q[x_i, t] dD^t . \tag{15.11}$$

By virtue of (3.20), the inverse Jacobi determinant obeys

$$\frac{\partial J^s_t}{\partial t} = -J^s_t \frac{\partial u_k}{\partial x_k} . \tag{15.12}$$

Hence $S_L\{q(t)\}$ is biased towards regions with a history of *convergence*, where drifters tend to accumulate. As pointed out by Middleton and Garrett (1986), for planar motions in a rotating reference frame, these are regions with a history of cyclonic relative vorticity ω,[1] in the quasigeostropic approximation ($|\omega| \ll |f|$). Middleton and Garrett speculate that the sampling bias towards cyclonic regions may be sufficient to yield a Lagrangian spatial average estimate of the polarization $C^A(\tau)$ having a sign opposite to that of an Eulerian spatial average.

Icebergs drifting over the Labrador Shelf provide Garrett et al. (1985) and Middleton and Garrett (1986) with a novel set of Lagrangian data: 224 tracks, consisting of one-hour positional fixes by radar located on the GUDRID oil well at (54°54′31″ N, 55°52′32″ W) in 299 m of water. *"Icebergs typically 35 m in draft and up to 50 km from GUDRID were radar tracked during 87 days following July 10, 1974"* (Middleton and Garrett, 1986). Four tracks are shown here in Figure 15.1. Note the clockwise precession in two of the tracks. Tidal currents are identified in the tracks by harmonic analysis at the dominant

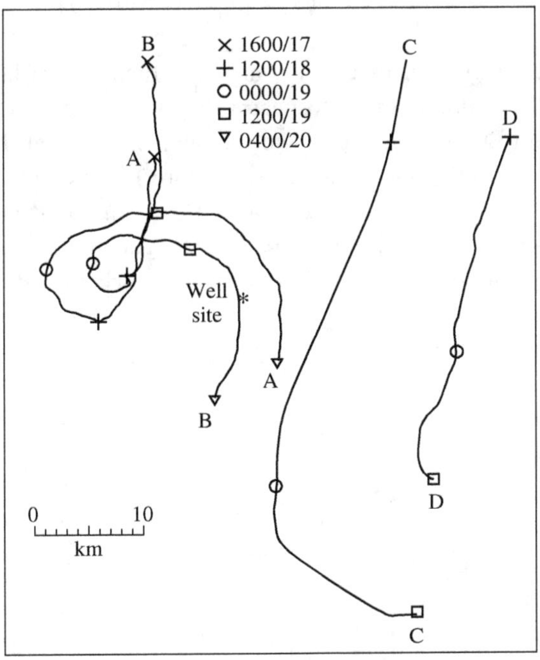

Figure 15.1 Four iceberg trajectories observed from GUDRID. Symbols show positions at hour and date in July 1974, after Garrett et al. (1985).

[1] That is, relative Cauchy invariant ϖ having the same sign as the Coriolis parameter f; see (3.60), (7.42), and consider perturbations $\delta\varpi$, δJ_t^s.

astronomical constituents (frequencies) and then removed. Drift owing to wind is identified by linear regression against local wind measurements, presumably at GUDRID, and then also removed. The resulting time series of velocity are statistically homogeneous and isotropic in space, and stationary in time, within a 95% confidence interval for all data. The Lagrangian velocity correlation functions $R_{uu}(\tau)$, $R_{vv}(\tau)$ and polarization $R^A(\tau)$, that is, the normalized covariances $C_{uu}(\tau)$, $C_{vv}(\tau)$ and normalized polarization $C^A(\tau)$ (for example, $R_{uu}(\tau) \equiv C_{uu}(\tau)/C_{uu}(0)$), are shown here in Figure 15.2. Isotropy is evident, as is clockwise polarization with a period between 10 and 20 hours. Lagrangian frequency spectra may be defined by

$$\Phi_{uu}(\sigma) = \int_{-\infty}^{\infty} R_{uu}(\tau) e^{-i\sigma\tau} d\tau , \tag{15.13}$$

etc. These may be combined as

$$D_{\pm}(\sigma) = \frac{1}{2}(\Phi_{uu}(\sigma) + \Phi_{vv}(\sigma)) \pm i\Phi^A(\sigma) . \tag{15.14}$$

Note that the autocovariances C_{uu}, C_{vv} are even functions of lag τ, while the polarization C^A is odd, so the autospectra Φ_{uu}, Φ_{vv} are real while the polarization spectrum Φ^A is imaginary. Accordingly, $D_{\pm}$ are real. Middleton and Garrett (1986) plot $\sigma D_{\pm}$ *versus* $\log_{10}\sigma$; see Figure 15.3. Note that $d(\log_{10}\sigma) \propto \sigma^{-1} d\sigma$, thus this plot has the same area as $D_{\pm}$ *versus* σ. If the polarization is anticlockwise (clockwise), then $D_+ > (<) D_-$. As already indicated by the sense and period of polarization in Figure 15.2, the spectra indicate clockwise

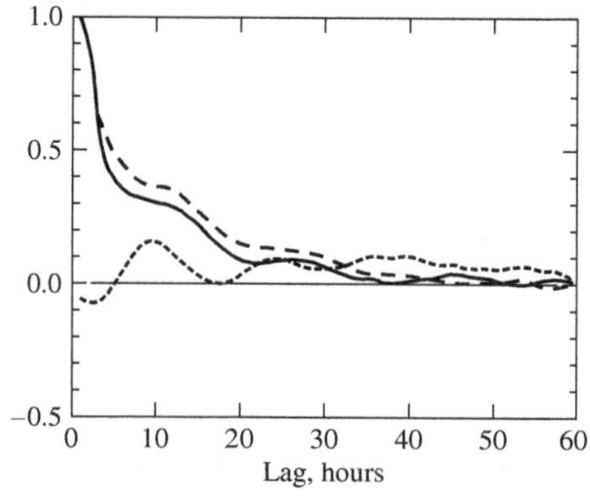

Figure 15.2 The Lagrangian velocity autocorrelations. *Solid line:* $R_{uu}(\tau)$, *dashed line:* $R_{vv}(\tau)$, *dotted line:* polarization $R^A(\tau)$; after Middleton and Garrett (1986).

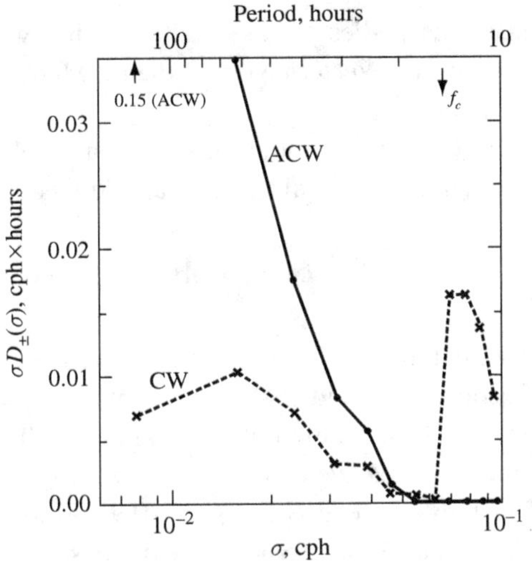

Figure 15.3 Lagrangian velocity spectrum $\sigma D_{\pm}$ *versus* $\log_{10}\sigma$. *Solid line:* σD_{+}, *dashed line:* σD_{-} ; after Middleton and Garrett (1986).

rotation with the inertial frequency $f = 2\Omega \sin\theta = (14.6)^{-1}$ cycles per hour at latitude 54° N. The spectra also indicate anticlockwise rotation of the velocity vectors below $(25)^{-1}$cph.

Exercise 15.3 Inertial oscillations in the Lagrangian velocity field satisfy

$$\frac{\partial u}{\partial t} - fv = 0\,, \tag{15.15}$$

$$\frac{\partial v}{\partial t} + fv = 0\,. \tag{15.16}$$

Show that the polarization is circular, and anticlockwise (clockwise) in the southern (northern) hemisphere. □

15.1.2 Diffusivities from floats

The estimation of Taylor diffusivities using float data is discussed in a pioneering paper by Freeland, Rhines and Rossby (1975). The data, obtained during the Mid-Ocean Dynamics Experiment (MODE) with 20 SOFAR floats ballasted to 1500 m, are described by Rossby, Voorhis and Webb (1975). The variability of the most simple statistics for kinetic energy prompt Rossby et al. to remark that "we should pay closer attention to the frequently made assumption that the scale of variation of the (second order) eddy statistics is large compared to the eddies themselves". In other words, MODE was conducted in a region of the Western North Atlantic which is not even locally statistically homogeneous.

Freeland et al. (1975) make a more detailed statistical analysis. Integral time-scales for the Lagrangian zonal velocity are defined as

$$T_u \equiv \int_0^\infty R_{uu}(\tau)d\tau , \tag{15.17}$$

for example. Thus a float time series for u, of duration T, possesses about $N = T/T_u$ degrees of freedom. Freeland et al. count a total of $N = 307$ independent degrees of freedom for all floats more than 55 km apart. They infer that estimates of, for example, the mean velocity component $E\{u\}$ have standard errors of $\sqrt{C_{uu}(0)/N}$; thus they arrive at $E\{u, v\} = (-0.9, -0.3) \pm (0.3, 0.3)\, cm\, s^{-1}$ (95% confidence interval, or two standard deviations).[2] They conclude that "the mean drift of the floats in the MODE region is certainly towards the west, but the north/south component is not significantly different from zero at the 95% confidence level". Freeland et al. make detailed estimates of 'chance' Eulerian flow statistics with the float data, but remark that "the experiment provided a generous number of nearly Lagrangian trajectories, and this aspect is of value far beyond the use of SOFAR floats as drifting current meters. The results are relevant to both tracer distributions and to dynamics".

Velocity–covariances $C_{uu}(\tau)$, etc., computed according to (15.4) and averaged over eight floats that had been tracked for 201–328 days, are shown in Figure 15.4. The parabolae which osculate C_{uu}, C_{vv} at zero lag intercept

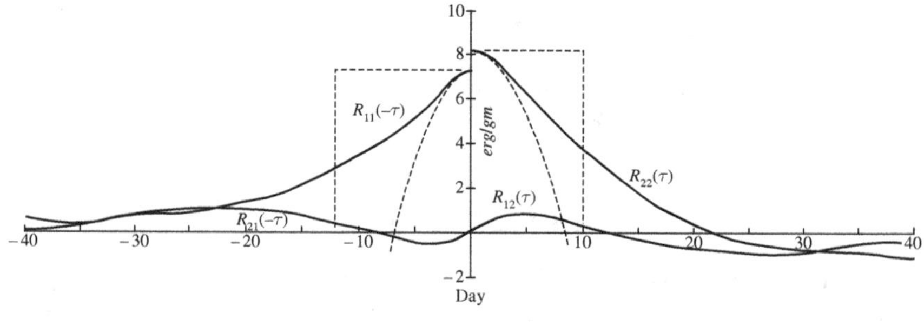

Figure 15.4 Lagrangian autocovariance tensor averaged over eight realization, after Freeland et al. (1975)

[2] For a beautiful explanation of such sample statistics for time series, see Lumley and Panofsky (1964), Part 1.B

the abscissa at 6.8 days and 8.4 days, while the integral time-scales T_u, T_v are 12.3 days and 10.1 days, respectively. Note that, for example, T_u is the base of the rectangle having the area under C_{uu}. Recalling from (10.11) that the Taylor diffusivity tensor is defined by

$$\kappa_{uu} \equiv \int_0^\infty C_{uu}(\tau)d\tau \tag{15.18}$$

etc., Freeland et al. arrive at the values $(7.8,7.1)\times 10^6 \mathrm{cm}^2\mathrm{s}^{-1}$ for κ_{uu}, κ_{vv}, respectively, which values they remark are "rather small compared with classical estimates of 1×10^8 to 4×10^8 $\mathrm{cm}^2\,\mathrm{s}^{-1}$ (Sverdrup, Johnson and Fleming, 1942) for the large scale ocean".

The kinematical identities (10.10) and (10.11) for the covariance of displacement of a single particle may be combined into

$$\mathrm{cov}\{X_i(a_k, s|t), X_j(a_k, s|t)\} = 2\int_s^t \int_s^v \mathrm{cov}\{u_i(a_k, s|v), u_j(a_k, s|w)\}^S dw dv\,, \tag{15.19}$$

where the superscript S denotes the symmetric part of the tensor. This is no more than a kinematic identity when the covariances are true expectation values; nevertheless, it has become known as "Taylor's theorem." For stationary turbulence it reduces to

$$\mathrm{cov}\{X_i(a_k, s|t), X_j(a_k, s|t)\} = 2\int_0^{t-s} \int_0^v C^S_{u_i u_j}(a_k|\tau)d\tau dv\,. \tag{15.20}$$

The single-particle velocity covariance in (15.20) is independent of the labeling time s, and may be computed by a time average as in (15.4). If the turbulence is homogeneous, then the displacement covariance is independent of the labeling position a_k, and so may be estimated by a space average as in (15.9). That is, a float average approximates an ensemble expectation value. Freeland et al. test these hypotheses by averaging the displacement products over 43 float tracks without regard to the different labeling times, and by integrating the sample covariances shown in Figure 15.4 (averaged in time, then averaged over eight floats). Their results are shown in Figure 15.5 in the form of rms displacement about the center of mass of the floats, which has the velocity $(-0.9, -0.3)\,\mathrm{cm\,s}^{-1}$. Inverse parabolic envelopes of displacement, depending on time like $\sqrt{t-s}$, would indicate a cloud of floats spreading out

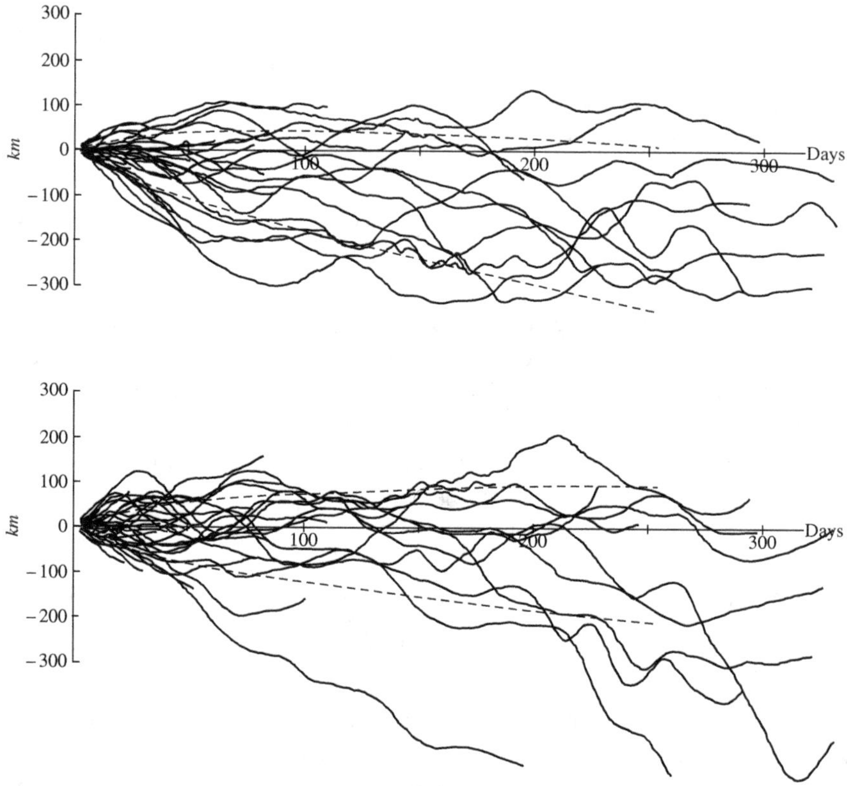

Figure 15.5 Displacement of floats in E/E and N/S directions, *upper panel* and *lower panel* respectively, *versus* time after launch of each float. *Dashed lines* show the expected dispersion (standard deviation about the center of mass) computed by integrating the autocovariance functions and adding in the mean flow; after Freeland et al. (1975)

about its center of mass by simple diffusion. As Freeland et al. stress, the most striking feature is the cessation of diffusion after 100 days; this would almost by definition be consistent with the influence of wave-like motion rather than turbulence. There should therefore be deep negative lobes in the velocity covariances, yet none is evident in Figure 15.4. For small times, the displacement covariance grows like

$$\mathrm{cov}\{X_i(a_k, s|t), X_j(a_k, s|t)\} \approx \mathrm{cov}\{u_i, u_j\}^S (t-s)^2 \tag{15.21}$$

where the velocity covariance is approximated as a time average at zero lag, for eight floats. Estimates of the standard deviations of displacement,

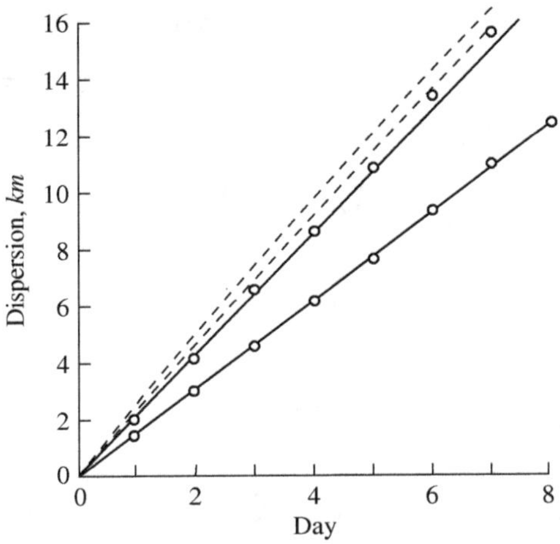

Figure 15.6 *Bold lines and circles*: the E/W and N/S measured dispersion. *Dashed lines*: the E/W and N/S dispersion predicted from the autocovariance functions of Figure 15.4. For each pair of lines the lower curve is the E/W dispersion and the upper curve the N/S dispersion; after Freeland et al., (1975).

from the lhs of (15.21) using the 43 float tracks, and also from the rhs, are shown in Figure 15.6. The squares of the line slopes for the lhs (*solid lines*) are $(3.2, 6.6)$ cm^2 s^{-2}; the squared slopes for the rhs (*dashed lines*) are $(7.3, 8.2)$ cm^2 s^{-2}. Freeland et al. speculate that the discrepancy owes to nonstationarity of the float velocities over 200–300 days. Since float velocities have a decorrelation time of 10 days, the authors accordingly reinitialize the calculation of the lhs of (15.21) and recompute the float-averaged displacement covariance every 30 days, yielding several independent estimates of the velocity variance. These estimates have means of $(3.9, 6.7)$ cm^2 s^{-2}, with standard deviations of $(0.6, 1.5)$ cm^2 s^{-2}. The authors remark: "Evidently non stationarity may explain the discrepancy in (the meridional component) but certainly not in (the zonal component)". This leaves inhomogeneity as the villain; in that case Taylor's theorem is, as Freeland et al. state, moot since there is then no pragmatic way to verify it.

Frequency spectra of Lagrangian velocities for a single long float track are shown in Figure 15.7. Freeland et al. note the energy-containing range at periods greater than 30 days, the steep slope of approximately (frequency)$^{-4}$ between 30 days and 10 days, and the peak near the inertial period (around 1 day). The noise in the range 10 days to 1 day is attributed to the SOFAR ranging system. The latter is arranged generally north–south of the MODE

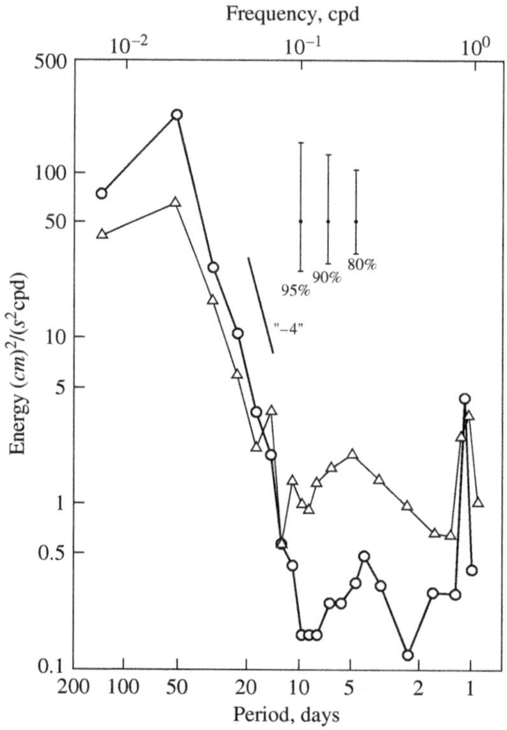

Figure 15.7 Lagrangian frequency spectra of N/S and E/W velocities, indicated by ○ and △, respectively. The average number of degrees of freedom for each estimate is 10; after Freeland et al. (1975).

region and is more accurate at determining transverse velocities, hence the greater noise level in the north–south component

Exercise 15.4 Assume that within the "−4" frequency range in Figure 15.7, the turbulence is cascading enstrophy from low to high wavenumbers at the constant rate λ. Can the slope of the frequency spectrum of velocity be inferred by dimensional analysis? In what sense might the assumption have any meaning? □

Drifters equipped with bio-optical sensors are now providing Lagrangian time series of observations of biologically important parameters such as temperature (Sea Surface Temperature, or SST), fluorescence, downwelling irradiance, upwelling irradiance and beam attenuation (Abbott et al., 1995). These parameters reflect changes in phytoplankton species composition, which in turn are influenced by changes in local ocean circulation. Lagrangian variance spectra reveal that the bio-optical parameters fluctuate on both diel and semidiurnal time-scales, perhaps associated with solar variation and semidiurnal tide, respectively (Abbott et al., 1995). As those authors remark, such signals must be carefully identified and eliminated in order to isolate genuine biological activity. Abbott and Letelier (1998) derive chlorophyll concentration from drifter measurements of bio-optical parameters in the California Current; examples of Lagrangian autocorrelation functions for chlorophyll and temperature are show in Figure 15.8. A decorrelation time is defined as the lag at which the correlation equals the 95% confidence interval. These scales are found to be six days for SST and four days for chlorophyll in the region between 200 and 400 km offshore. In the region more than 400 km offshore, the decorrelation time-scale is 7 days for SST but is as small as 2.5 days for chlorophyll (see Figure 15.9), prompting Abbott and Letelier to comment that "the processes regulating the distribution of temperature and chlorophyll are similar in the nearshore region and significantly different offshore". They cite Bennett and Denman (1985) who argue that the only mechanism which could cause biological patterns to deviate from the spatial patterns of mesoscale processes would be spatial heterogeneity in net growth rates. To the extent that drifters sample spatial pattern, such deviation and

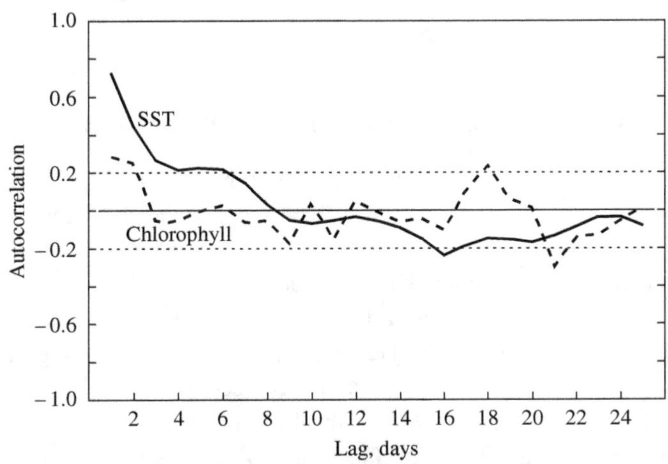

Figure 15.8 An example of the temporal autocorrelation function of SST and chlorophyll. *Dashed lines* represent 95% confidence intervals based on the assumption that the input series is a white-noise process; after Abbott and Letelier (1998).

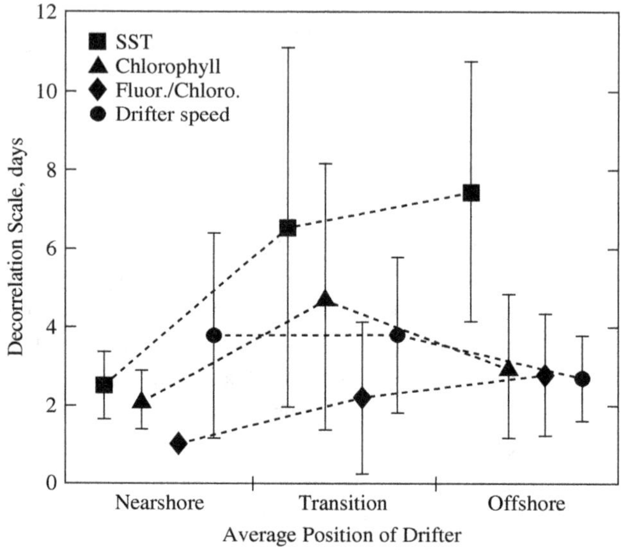

Figure 15.9 Average decorrelation scales for SST, chlorophyll, fluorescence/chlorophyll, and drifter speed. Data are plotted as a function of the average distance offshore of the drifter tracks. The error bars represent ±1 standard deviation; after Abbott and Letelier (1998)

hence such spatial variability in net growth rate should be detectable in drifter data.

15.2 Assimilation: the single particle

Oceanic and atmospheric data assimilation schemes have been developed almost exclusively for Eulerian fluid dynamical models. Moreover, almost all data types are Eulerian: they are collected by sensors fixed in space, or moving on a precisely controlled path such as a satellite orbit. Data from ships under way and from aircraft are complicated exceptions, of course; they will be considered shortly. Development of assimilation algorithms for Eulerian models and data is very advanced: see for example monographs (Bennett, 1992, 2002; Daley, 1991; Wunsch, 1996), review articles (Anderson, Sheinbaum and Haines, 1996; Malanotte-Rizzoli, 1996; Fukumori, 2001) and major sequences of journal articles (Rabier et al., 2000; Mahfouf and Rabier, 2000; Klinker et al., 2000). Operational experience is far greater in meteorology than in oceanography; owing to a dearth of real-time ocean data, the oceanic emphasis has been largely on scientific testing of climate models (e.g., Bennett et al., 1998, 2000, 2005; Keppenne, 2000; Stammer et al., 2003[3]), but see also Evensen (1994)[4] for an ocean assimilation and forecasting scheme currently in operation.

The most extensively assimilated oceanic data have been Eulerian measurements of the mass field, that is, subsurface hydrography (Stammer et al., 2003) and surface altimetry (Fukumori, 2001). The Tropical Atmosphere Ocean or TAO array in the Pacific Ocean (McPhaden, 1995) provides Eulerian measurements of subsurface temperature near the equator, and subsurface currents on the equator. The temperature data have been assimilated by, for example, Bennett et al. (1998, 2000, 2005). The extensive measurements of near-surface temperature from Volunteer Observing Ships (VOS) using Expendable Bathythermographs (XBTs) have been assimilated by Keppenne (2000), for example. The numerous surface drifter data and far less numerous float data have attracted less attention. Both drifters and floats have almost invariably been treated as Eulerian measurements resembling VOS data. The first attempts at assimilating Lagrangian data into Lagrangian models will be outlined in the following subsections. It is necessary first to examine Lagrangian measurement functionals, with particular emphasis on their linearity or otherwise, since nonlinearity greatly complicates assimilation algorithms and measurement error statistics.

15.2.1 Lagrangian measurement functionals

The dependent variable or *state* in an Eulerian model is in general a multivariate U_p, $1 \le p \le K$, with components such as fluid velocity u_i, pressure P, density ρ, temperature T and other tracers such as relative humidity or salinity, etc. These components are all fields over space and time, thus $U_p = U_p[x_j, t]$. Consider for example a thermometer moored at a fixed location x_j^d, taking a measurement of the real ocean state at time t^d. The thermometer will record the datum T^d. This real recording may be mimicked mathematically with a functional acting on the model state:

$$U_p \to T[x_j^d, t^d]\,. \tag{15.22}$$

Note that the state on the lhs of (15.22) is a function, in fact a multivariate field over the four dimensions of space and time, while the particular field value on the rhs is a single number. The functional (15.22) is clearly linear: if $U_p^{(1)}$ and $U_p^{(2)}$ are two such functions which are mapped into the single

[3] http://www.ecco–group.org/
[4] http://topaz.nersc.no/

numbers $T^{(1)}$ and $T^{(2)}$, respectively, then for any constants $c^{(1)}$ and $c^{(2)}$ the linear combination of functions is mapped as

$$c^{(1)}U_p^{(1)} + c^{(2)}U_p^{(2)} \to c^{(1)}T^{(1)} + c^{(2)}T^{(2)} . \tag{15.23}$$

An important counterexample is provided by atmospheric radiative intensity R as measured by an earth satellite: assuming that absolute temperature T is a state variable in an atmospheric model, Stefan's law has $R \propto T^4$.

Consider now the temperature T measured by a deep float or a high-altitude balloon. The measurement functional is

$$U_p \to T(a_k^d, s|t^d) = T[X_j(a_k^d, s|t^d), t^d] , \tag{15.24}$$

where (a_k^d, s) labels the float or balloon. This is evidently a nonlinear functional acting on the Eulerian state, since of course the path of fluid particles depends upon the state of the current or wind. It is conventional to regard the locations of the measurements as fixed during the assimilation; this would be strictly correct if admissable variations of the velocity field were constrained to be exactly consistent with the observed path of the float or balloon, but such a constraint seems never to be imposed. The difficulty is obviated by expressing the model in Lagrangian variables. The state is then a multivariate field over labels and time, of the form

$$U_p = U_p(a_k, s|t) , \tag{15.25}$$

and the float or balloon temperature datum is mimicked by a *linear* measurement functional:

$$U_p \to T(a_k^d, s|t^d) . \tag{15.26}$$

Consider the Eulerian temperature measurement $T[x_i^d, t^d]$, which is *defined* in Chapter 1 by

$$T[x_i^d, t^d] \equiv T(x_i^d, t^d|t^d) . \tag{15.27}$$

That is, the measurement by a thermometer moored at position x_i^d at time t^d is the same as that of a thermometer on a float (or drifter or balloon) released at that same place, at that same time. It would appear that, as a special case of (15.26), Eulerian measurement functionals are linear in Lagrangian variables for all quantities having linear Lagrangian measurement functionals. However this is not useful, since the Lagrangian description of the state is labeled at one time s. In terms of the labels at that time, the Eulerian temperature data are

$$T[x_i^d, t^d] = T\left(X_j(x_i^d, t^d|s), s|t^d\right), \tag{15.28}$$

which depends not only upon the temperature field in Lagrangian variables, but also upon the Lagrangian velocity field. Equally, there is no eliminating the nonlinearity of $T[X_j^d, t^d]$ in (15.24) if the path X_j^d is properly allowed to vary with the flow, or if the path is regarded as an imperfect measurement during assimilation. Radiation is a nonlinear function of state in both the Eulerian and Lagrangian formulations. Temperature may be replaced with radiation as a state variable, thereby rendering that functional linear, but then thermodynamic conservation laws such as (3.21) acquire further nonlinearity since $\mathfrak{E} \propto R^{\frac{1}{4}}$.

As already mentioned, data collected from self-propelled platforms (watercraft or aircraft, manned or otherwise) represent a mixture of Eulerian and Lagrangian variables. Let the symbols u, v, w be recycled yet again, with u_i being the fluid velocity, v_i the velocity of self-propulsion or velocity in a resting fluid, w_i the true or total velocity of the craft, and $Z_i(a_j, s|t)$ the true path of the craft. Then

$$\frac{\partial}{\partial t} Z_i(a_j, s|t) = w_i(a_j, s|t) = u_i[Z_k(a_j, s|t), t] + v_i(a_j, s|t), \tag{15.29}$$

for a craft launched at position a_j at time s. In particular, note that

$$Z_i(a_j, s|t) \neq X_i(a_j, s|t) + Y_i(a_j, s|t), \tag{15.30}$$

where $X_i(a_j, s|t)$ is the path of a fluid particle, and $Y_i(a_j, s|t)$ is the path of the craft relative to the fluid. Moroever, for any quantity q measured from the craft,

$$q[Z_i(a_j, s|t), t] \neq q[X_i(a_j, s|t), t] = q(a_i, s|t). \tag{15.31}$$

It may be possible to determine the true path by an earth positioning system such as celestial navigation, RAFOS/SOFAR, LORAN or the Global Positioning System (GPS). The relative path Y_i may be determined by dead reckoning, that is, by integrating the relative velocity v_i of the craft. However, the path of a single fluid parcel cannot be inferred, since the craft is continuously moving from one fluid parcel to another. The entanglement of velocities and parcels in (15.29) is nonlinear, and measurements from the craft would seem to be nonlinear functionals for states expressed as fields over either Eulerian or Lagrangian variables. This is the case also for state variations when the relative path is known. However, if the true position Z_i of the craft at time t *is* known, then measurements from it are of the form $q[Z_i^d, t^d]$, which is Eulerian.

In the absence of a knowledge of the true path, the measurement functionals are inevitably mixed. If the relative velocity of the craft is insignificant, then measurements from it are approximately Lagrangian; if the fluid velocity is insignificant, then measurements are Eulerian in the sense that their positioning and timing are at the disposal of the observer. In the first case, these measurements are linear only in Lagrangian variables (that is, the measurement functionals are linear only with respect to states expressed in Lagrangian variables, and only if the measured quantity depends linearly on the state components); in the second case the measurements are linear in Eulerian variables.

15.2.2 Lagrangian assimilation: first steps

It is fitting that perhaps the first study of assimilation of Lagrangian data should have been conducted at the University of Rhode Island, which has been a leader in developing floats and using them to investigate ocean circulation.[5] Carter (1989) proposes that the isopycnal velocity and depth of an isopycnal float should obey the single layer, reduced gravity, shallow-water equations:

$$\frac{\partial u}{\partial t}+u\frac{\partial u}{\partial x}+v\frac{\partial u}{\partial y}-fv=-g'\frac{\partial h}{\partial x}, \tag{15.32}$$

$$\frac{\partial v}{\partial t}+u\frac{\partial v}{\partial x}+v\frac{\partial v}{\partial y}+fu=-g'\frac{\partial h}{\partial y}, \tag{15.33}$$

$$\frac{\partial h}{\partial t}+\frac{\partial (uh)}{\partial x}+\frac{\partial (vh)}{\partial y}=0, \tag{15.34}$$

where all derivatives are Eulerian and the coordinates (x, y) are Cartesian. The active layer current is $\big(u[x, y, t], v[x, y, t]\big)$ which Carter identifies with the isopycnal velocity of the float, and $h[x, y, t]$ is the active layer thickness which Carter identifies with the float depth. Meanwhile f is the Coriolis parameter on the β plane, and g' the reduced value of gravity (see, e.g., Gill, 1982; Pedlosky, 1987). The data consist of time sequences of labeled velocity components $\big(u(a, b, s|t), v(a, b, s|t)\big)$ and labeled depth $h(a, b, s|t)$ where the label is the position (a, b) of the float at time s. The plane components

[5] `http://mail.po.gso.uri.edu/rafos/index.html`

$\big(X(a, b, s|t), Y(a, b, s|t)\big)$ of subsequent float position are also known. The state may be organized as a triple $\mathbf{U} = (u, v, h)$, which is a field over two-dimensional space (x, y) and over time t. Space is discretized, and the spatial nodes are enumerated from 1 to N. Thus at any time the state becomes a long vector of the form $(\mathbf{U}_1, \ldots, \mathbf{U}_n, \ldots, \mathbf{U}_N)$. The measurement functionals for the three data at the nth spatial node, at any time, constitute a matrix of the form $(\mathsf{0}, \ldots, \mathsf{I}, \ldots, \mathsf{0})$, where $\mathsf{0}$ is the 3×3 zero matrix and I is the 3×3 unit matrix. Carter remarks that the measurement functional for each observation triple need not be stored in full; only a single pointer need be stored. Carter chooses to assimilate the data sequentially in time, using the Kalman filter (see, e.g., Gelb, 1974; Ghil et al., 1981; Miller, 1986; Bennett, 2002). The filter forecasts the spatial covariance of the error in the forecast of the state. Computation and management of the $3N \times 3N$ error covariance matrix is greatly reduced by Carter, who argues that the covariance is negligible at a finite distance from a node. Sophisticated data structures track the finite regions of influence of the floats. Carter shows results for a one-dimensional linear model and a two-dimensional nonlinear model; the dynamical linearization essential for the Kalman filter algorithm is not described. This highly innovative merging of float data with a model is neverthless strictly Eulerian. First, the dynamics (15.32)–(15.34) are expressed in Eulerian variables. Second, the float paths are exactly those observed, and so are not consistent with the Kalman filter forecast of the velocity field.

A mixed assimilation of tropical Pacific drifter tracks into the nonlinear shallow-water equations on the sphere may be found in Kamachi and O'Brien (1995). The equations of motion are Eulerian, but the measurement functionals are Lagrangian and therefore nonlinear with respect to the Eulerian representation of the ocean state. The drifter tracks are not considered to have been perfectly observed, and the model tracks are not required to fit the observations exactly. The state estimate is an exact solution of the numerically approximated equations of motion, and is a weighted least squares best fit to the tracks over "smoothing intervals" of either three months or one year. Controls include the initial state of the model, and various parameters in the dynamics. The fit is found by descending the gradient of the fitting criterion, or penalty functional, with respect to the controls. The gradient is computed by integration of the adjoint equations backward in time (Talagrand and Courtier, 1987; Bennett, 2002). Observed and fitted paths for one assimilation are shown in Figure 15.10.

Another innovative approach to assimilating Lagrangian data is owed to Ide, Kuznetsov and Jones (2002), who assume planar incompressible, irrotational

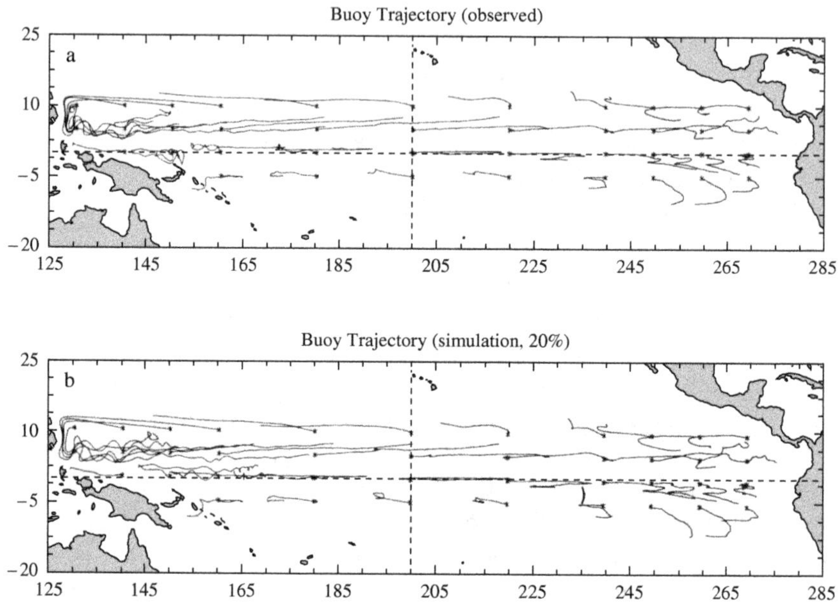

Figure 15.10 Examples of drifting buoy trajectories of (a) observations and (b) simulations, after Kamachi and O'Brien (1995).

flow in the form of N_F vortices. The complex Eulerian velocity field $w \equiv u - iv$ at the plane position $z \equiv x + iy$ satisfies

$$w[z, t] = \frac{1}{2\pi i} \sum_{n=1}^{N_F} \frac{\Gamma_n}{z - z_n}, \tag{15.35}$$

save at the vortex cores $z = z_n$, $1 \le n \le N_F$. The real-valued circulation around the nth core is Γ_n. The path of the nth core satisfies

$$\frac{\partial z_n}{\partial t} = \frac{i}{2\pi} \sum_{m=1, n\neq m}^{N_F} \frac{\Gamma_m}{\bar{z}_n - \bar{z}_m} + \eta_{F_m}, \tag{15.36}$$

for $1 \le n \le N_F$. The overbar in (15.36) denotes the complex conjugate. The additional core velocity $\eta_{F_m}(t)$ is white noise intended to represent unresolved dynamical processes. Introducing N_D drifters into the flow, their complex paths $Z_n = X_n + iY_n$ obey

$$\frac{\partial Z_n}{\partial t} = \frac{i}{2\pi} \sum_{m=1}^{N_F} \frac{\Gamma_m}{\bar{Z}_n - \bar{z}_m} + \eta_{D_n}, \tag{15.37}$$

for $1 \leq n \leq N_D$. The additional core velocity $\eta_{D_m}(t)$ is white noise intended to represent drifter diffusion. Power series expansions of the rational functions in (15.36) and (15.37) about known core paths and known drifter paths yield, at first order, linearized forms of these "augmented dynamics" for the "augmented state" (z_n, Z_m), thereby enabling the Kalman filter algorithm (Gelb, 1974) for sequential assimilation of drifter positions into the flow (15.35). Note that at any time t, the state is characterized by the positions of the N_F vortex cores and the N_D drifters. This highly ingenious assimilation scheme is of mixed form: the forecast drifter paths are consistent with the forecast vortex paths and hence with the forecast velocity field, but the flow is characterized by the incompressible, irrotational, Eulerian form (15.35).

Exercise 15.5 Devise a Kalman filter for the incompressible, rotational, Ptolemaic vortices of Section 7.5, and the data of Ide et al. (2002). Is this assimilation problem purely Lagrangian? Is state augmentation necessary? Is linearization necessary? How should subgridscale noise be introduced? □

A general-purpose Lagrangian assimilation scheme is being developed by Mead and Bennett (2001), Mead (2004, 2005). The dynamics are the shallow-water equations on the surface of a rotating sphere[6]. Admitting rotation and neglecting radial velocity, the conservation equations (4.15) and (4.16) for momentum become

$$\frac{\partial^2 \Phi}{\partial t^2} - \left(2\frac{\partial \Phi}{\partial t} + 2\Omega\right)\frac{\partial \Theta}{\partial t}\tan\Theta = -\frac{g'}{(R\cos\Theta)^2 J_s^t}\frac{\partial(h,\Theta)}{\partial(\alpha,\beta)}, \tag{15.38}$$

$$\frac{\partial^2 \Theta}{\partial t^2} + \left(\frac{\partial \Phi}{\partial t} + 2\Omega\right)\frac{\partial \Phi}{\partial t}\sin\Theta\cos\Theta = -\frac{g'}{R^2 J_s^t}\frac{\partial(\Phi,h)}{\partial(\alpha,\beta)}. \tag{15.39}$$

Conservation of volume becomes

$$\frac{\partial(h\cos\Theta J_s^t)}{\partial t} = 0, \tag{15.40}$$

where Ω is the constant rate of rotation about the north pole ($\Theta = \pi/2$), g' is a reduced value for gravity and R is the earth's radius, while (α, β) are the labels at the release time s. The Jacobi determinant is

$$J_s^t \equiv \frac{\partial(\Phi,\Theta)}{\partial(\alpha,\beta)}. \tag{15.41}$$

[6] Those authors prefer the notation (λ, θ) to the notation (Φ, Θ) used here in Section 4.2 for the longitude and latitude of a fluid particle.

Exercise 15.6 Establish the law of conservation of volume in the form (15.40). Show that the cosine factor is the value of a determinant. □

These equations are transcendentally nonlinear, owing to the trigonometric functions of the particle latitude Θ. Other nonlinearities owe to centrifugal effects in curvilinear coordinates, to the variability of the Coriolis parameter $2\Omega\sin\Theta$, to the nonlinearity of the pressure gradient in Lagrangian coordinates and to the conservation of fluid volume within a layer of variable thickness. On the other hand, data for float position and depth are associated with linear measurement functionals.

It is clear that these equations define a well-posed initial boundary value problem in rigidly bounded domains; indeed, the boundary conditions are quite simple: particles initially on the boundary may subsequently move around it, but remain on it indefinitely. There is no need to specify the height h on the boundary. Moreover, Bennett and Chua (1999) argue theoretically and demonstrate computationally that the forward problem for these equations is well posed in open domains that are comoving, or moving with the flow; see also Section 9.5. Such a computational domain is particularly suited to analyzing the flow in the neighborhood of a cluster of a floats. It may not be suited to a requirement of long-term analysis of flow in a geographically fixed ocean region. In that case, floats may be a poor choice of observing system. The shallow-water equations also lead to well-posed problems in fixed open regions, provided both the sign and magnitude of the local Froude number $u_i\hat{n}_i/\sqrt{g'h}$ are taken into account when choosing both the number and form of the boundary conditions at an open boundary having unit outward normal $\hat{n}_i$. However, such considerations become impractical upon generalizing from the shallow-water equations to the hydrostatic primitive equations (Oliger and Sundström, 1978).

Exercise 15.7 Derive an integral inequality, analagous to (2.23) in Bennett and Chua (1999), for the total mechanical energy of the difference between two solutions of (15.38)–(15.41) in a comoving domain. Deduce the boundary conditions necessary for a unique solution to the mixed initial boundary value problem. Note that the argument does not require the equations of motion to be in Lagrangian form, although that form does simplify the definition of the comoving boundary. □

The intended application of the inviscid shallow water model (15.38)–(15.41) is the assimilation of float data collected in the subtropical North

Atlantic, with sufficient resolution for the mesoscale variability first detected by Swallow's floats (see Chapter 14). Explicit representations of turbulent transfers of momentum and volume may not be needed for simulation of mesoscale dynamics, especially as the intended application is variational assimilation of float data with the dynamics only as weak constraints. Yet there will be a turbulent transfer of enstrophy to high wavenumbers, leading to an accumulation at the highest resolved wavenumber. For a finite-difference analysis of the enstrophy cascade, see Bennett and Middleton (1983). The resulting instability can cause the height h to become negative, which besides being physically absurd renders the dynamics of gravity waves elliptic and thereby creates further ill-posedness. In short, a simple parametrization of turbulent dissipation must be included. A constant eddy viscosity is the simplest parametrization that is largely restricted to higher wavenumbers. The associated operators for the divergence of the rate-of-strain tensor are complicated enough in spherical polar coordinates.[7] The operators are inordinately complex when expressed in Lagrangian variables. Although great care is needed in order to compare numerical solutions with those in Eulerian spherical polar coordinates, the eventual purpose of simply absorbing the turbulent cascade justifies substantial simplifications such as "freezing" the latitude when expanding the operators. For details, see Mead and Bennett (2001), Mead (2004, 2005).

Unlike the Kalman filter algorithm for sequential estimation, fixed interval smoothing by variational means may be defined for nonlinear dynamics and measurement functionals. The hybrid scheme of Kamachi and O'Brien (1995) is an example. Extrema of the penalty functional satisfy the Euler–Lagrange conditions: see, for example, Courant and Hilbert (1953), Lanczos, 1966, Bennett and McIntosh (1982), Bennett (1992, 2002), etc. Efficient solution of these nonlinear equations requires iteration on linearized equations, but the iteration need not be based on the *tangent* linearization (Talagrand and Courtier 1987). As can be imagined, linearization of even the inviscid shallow water equation in Lagrangian variables on the sphere is especially intricate (Mead and Bennett, 2001; Mead, 2004, 2005). Diffusion operators, which are nonlinear in Lagrangian variables, only add to the misery. All that can be said thus far is that tests with synthetic flows and data typical of North Atlantic mesoscale variability are encouraging.

[7] See for example, Landau and Lifschitz (1959) but note that their coordinates are radial distance r, colatitude $(\pi/2-\theta)$ and longitude ϕ *in that order*.

16

Data analysis: particle clusters

16.1 Time series analysis: the particle pair

It is essential to begin with a review of the basic statistics of displacement of a single particle, as presented in Chapter 10.

Exercise 16.1 Show that the covariance of the components of the displacement vector satisfies the kinematic identity

$$\operatorname{cov}\{X_i(a_k, s|t), X_j(a_k, s|t)\} = \int_s^t \int_s^t \operatorname{cov}\{u_i(a_k, s|v), u_j(a_k, s|w)\} dw dv. \tag{16.1}$$

It may be recalled from Chapter 2 that if the multi-point, single-time Eulerian velocity field is statistically homogeneous and stationary, then the single-particle Lagrangian velocity is statistically homogeneous and its autocovariance at two times is time translation invariant. Show that under such conditions the displacement variance is independent of release position:

$$\operatorname{cov}\{X_i(a_k, s|t), X_i(a_k, s|t)\} = 2\int_0^{t-s} \int_0^{v} C^S_{u_i u_i}(w) dw dv, \tag{16.2}$$

and has the asymptote

$$\operatorname{cov}\{X_i, X_i\} \sim 2(t-s)\int_0^{\infty} C^S_{u_i u_i}(w) dw \tag{16.3}$$

as $t - s \rightarrow \infty$. □

Seemingly similar formulae hold for the separation of a pair of particles.

Exercise 16.2 Consider a pair of particles at time s, one at a_k and the other at $a_k + f_k$; see Figure 11.1 and Figure 11.2. Show that the covariance of

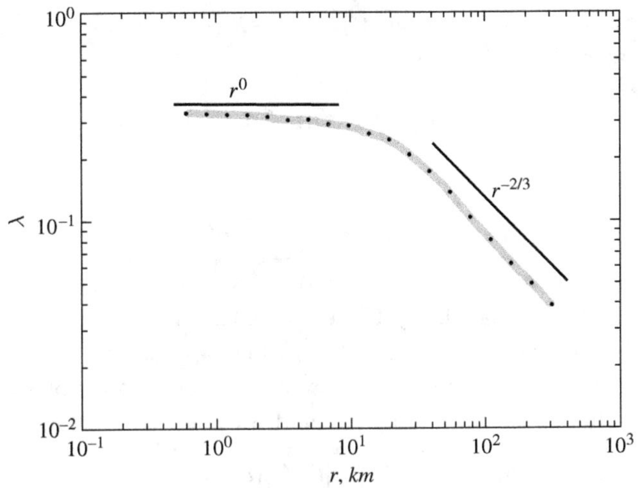

Figure 16.1 The finite scale Lyapunov exponent for a set of drifters in the Gulf of Mexico. The thick gray lines indicate the 95% confidence limits. The implied Lyupanov exponent is nearly constant at small scales, and decays as $r^{-2/3}$ at large scales; after LaCasce and Ohlmann (2003).

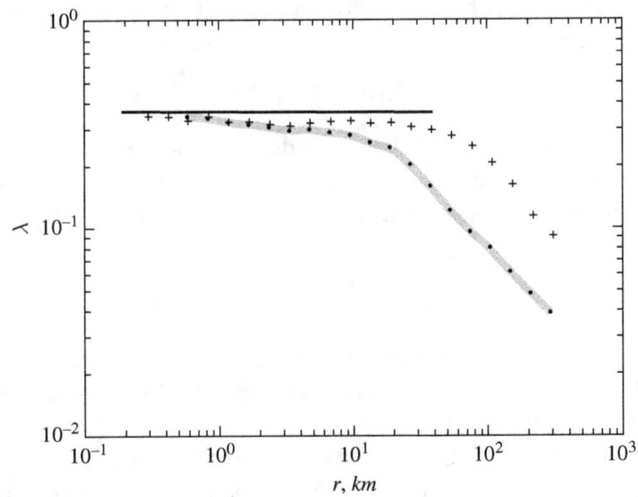

Figure 16.2 The finite scale Lyapunov exponents for the full set (.) and the subset (+) which have $r_0 \leq 1$ km. The Lyapunov exponent is nearly constant over a larger range of scales in the subset; after LaCasce and Ohlmann (2003).

their separation vector $r_i \equiv X_i(a_k+f_k, s|t) - X_i(a_k, s|t)$ at time t satisfies the kinematic identity

$$\text{cov}\{r_i, r_j\} = \int_s^t \int_s^t \text{cov}\{\mu_i(v), \mu_j(w)\} dv dw, \tag{16.4}$$

where

$$\mu_i(t) \equiv u_i(a_k+f_k, s|t) - u_i(a_k, s|t) \tag{16.5}$$

is the Lagrangian velocity of separation. The pair of spatial labels for μ_i is being suppressed for clarity alone; statistical homogeneity is not being assumed. For the case of the velocities of the two particles being asymptotically independent for large time, show that the covariance of separation is asymptotically the sum of the individual displacement covariances:

$$\begin{aligned} \text{cov}\{r_i, r_j\} \sim & \text{cov}\{X_i(a_k, s|t), X_j(a_k, s|t)\} \\ & + \text{cov}\{X_i(a_k+f_k, s|t), X_j(a_k+f_k, s|t)\} \end{aligned} \tag{16.6}$$

as $t-s \to \infty$. □

The separation vector, being the difference of two asymptotically independent and normally distributed displacement vectors, is asymptotically normal. If the Lagrangian single particle velocity covariance is homogeneous, then by (16.6) the separation covariance asymptotes to twice the displacement covariance:

$$\text{cov}\{r_i, r_j\} \sim 2\text{cov}\{X_i(a_k, s|t), X_j(a_k, s|t)\} \tag{16.7}$$

as $t-s \to \infty$. Indeed, for a homogeneous and stationary Eulerian velocity field, the separation velocity is homogeneous. And once the Lagrangian velocities of the pair have become uncorrelated, the separation velocity becomes stationary:

$$\text{cov}\{\mu_i(v)\mu_j(w)\} \sim 2C_{u_i u_j}(w-v) \tag{16.8}$$

for all $v-w$, as $v-s$ and $w-s \to \infty$.

The critical issue for sample estimation is that even if the Eulerian velocity field u_i is homogeneous and stationary, the velocity of pair separation μ_i defined in (16.5) is *not* stationary while the velocities of the two particles are correlated, as remarked in Section 2.3. Middleton (1980a,b) reviews some estimates of separation statistics for drifter data, with regard to this issue. As pointed out by Richardson (1926), it is first necessary to identify and eliminate deterministic velocity fields such as tides, seasonal circulation and steady general circulation.

The development of relative dispersion in Chapter 11 turns on the relative diffusivity ζ, which is the ensemble mean of the time rate of change of squared separation, the mean being taken over those pairs having the same separation r at that time; see (11.26). LaCasce and Ohlmann (2003) compute instead the finite scale Lyapunov exponent (FSLE), for surface drifters in the Gulf of Mexico. Consider a geometric progression of separations:

$$r^{(n+1)} = \alpha r^{(n)}, \tag{16.9}$$

where α is a constant. Let $T^{(n)}$ be the time interval during which the separation of a drifter pair increases from $r^{(n)}$ to $r^{(n+1)}$. Then the FSLE is defined to be

$$\lambda^{(n)} = \left\langle \frac{1}{T^{(n)}} \right\rangle \ln \alpha, \tag{16.10}$$

where the angle brackets denote the sample mean for all pairs having the separation $r^{(n)}$ at the beginning of the interval. These pairs need *not* have the same separation at the same universal time, so the sample sizes can be relatively large. LaCasce and Ohlmann (2003) plot $\lambda^{(n)}$ against $r^{(n)}$, for $\alpha = \sqrt{2}$; see Figure 16.1. The FSLE is approximately constant at 0.35 per day (e folding time of about 3 days) for separations smaller than 10 km, but falls off as $r^{-2/3}$ at larger separations. A conventional relative dispersion analysis of the same data suggests an e folding time of about 2 days, for separations as large as 50 km. The discrepancy in the time-scales is not as profound as in the separation scales; LaCasce and Ohlmann recompute the FSLE for those drifters for which $r^{(0)} \leq 1$ km. The value for the subset is much the same as for the full set, but is sustained up to 40–50 km; see Figure 16.2. Those authors argue that the drifters with initial separations in the range $1 \leq r \leq 10$ km must have had correlated initial separations and velocities, which condition would have impeded the onset of the separation range characterized by exponential growth statistics.

The use of Lyapunov exponents is representative of an emerging trend, away from statistical analyses of particle kinematics, towards analyses motivated by the concept of chaotic motion (see, e.g., Samelson and Wiggins, 2005). A chaotic analysis of rotary polarization awaits development.

16.2 Assimilation: particle clusters

16.2.1 Eulerian kinematical analysis

Clusters of three of more drifters or floats or balloons may be used to estimate spatial gradients of large-scale flow. Kirwan (1975) summarizes older results;

Fahrbach, Brockmann and Meincke (1986) apply the technique to equatorial North Atlantic drifters. The essence is simple: let (X_n, Y_n) for $1 \le n \le N$ be the plane coordinates of N drifters. Their centroid is at

$$(\overline{X}, \overline{Y}) \equiv N^{-1} \sum_{n=1}^{N} (X_n, Y_n). \tag{16.11}$$

Expanding the Eulerian velocity field about the centroid yields

$$u_n = \overline{u} + \overline{\frac{\partial u}{\partial x}}(X_n - \overline{X}) + \overline{\frac{\partial u}{\partial y}}(Y_n - \overline{Y}) + u_n'', \tag{16.12}$$

$$v_n = \overline{v} + \overline{\frac{\partial v}{\partial x}}(X_n - \overline{X}) + \overline{\frac{\partial v}{\partial y}}(Y_n - \overline{Y}) + v_n'', \tag{16.13}$$

for $1 \le n \le N$, where the first-order remainders (u_n'', v_n'') are attributed to relatively small-scale flow. Note that the overbars on the spatial gradients in (16.12) and (16.13) indicate evaluation at the centroid. The requirement that

$$(\overline{u_n}, \overline{v_n}) = (\overline{u}, \overline{v}) \tag{16.14}$$

yields $(\overline{u_n''}, \overline{v_n''}) = (0, 0)$. The drifter velocities (u_n, v_n) may be evaluated by time differencing the drifter tracks, hence (16.12) and (16.13) constitute $2N$ linear relations for the four spatial gradients at the centroid, with residuals (u_n'', v_n''). These linear relations are subject to the two linear constraints (16.14). If the cluster is a triple, $N = 3$, then there is a unique solution for the four spatial gradients, with vanishing residuals. If $N > 3$, then there is a solution which minimizes the residuals in a least-squares sense. Given the values of the four spatial gradients, plus the velocities of any $N - 3$ drifters, all $2N$ residuals may be reconstructed. Thus an unbiased estimate of the residual variance has $N - 3$ degrees of freedom (Sanderson, Pal and Goulding, 1988):

$$(\sigma_u^2, \sigma_v^2) = \frac{1}{N-3} \sum_{n=1}^{N} \left(u_n''^2, v_n''^2\right). \tag{16.15}$$

Fahrbach et al. (1986), following Okubo and Ebbesmeyer (1975), compute the cluster standard deviation of distances (l_x, l_y) from the centroid according to

$$(l_x^2, l_y^2) \equiv \frac{1}{N-1} \sum_{n=1}^{N} \left((X_n - \overline{X})^2, (Y_n - \overline{Y})^2\right), \tag{16.16}$$

and then estimate *relative* diffusivities as

$$(K_x, K_y) = c(l_x \sigma_u, l_y \sigma_v) \tag{16.17}$$

where $c \approx 0.1$. That is, the cluster standard deviations are identified with mixing lengths, and the residual variances with eddy velocity variances. The results of Fahrbach et al. suggest $K_{x,y} \sim l_{x,y}^{1.43}$. The exponent is not significantly different from the value of 4/3 consistent with an energy range, see (16.70), but is more likely indicative of absolute dispersion in a shear flow.

Exercise 16.3 (Saucier, 1955) By analogy with (3.19), show that the area A^t of a fluid patch evolves according to

$$\frac{dA^t}{dt} = \overline{\delta} A^t \tag{16.18}$$

where

$$\delta \equiv \frac{\partial u}{\partial x} + \frac{\partial v}{\partial y} \tag{16.19}$$

is the flow divergence, and the overbar denotes the area average. A triple of drifters define a triangular patch having an area given by half the magnitude of the vector product of any pair of sides, hence δ may be estimated as $\overline{\delta}$.

Consider a rotation of axes through $\pi/2$ anticlockwise. The velocity components transform according to $(u' = v,\ v' = -u)$. The relative vorticity $\zeta = v_x - u_y$ transforms into the divergence $\delta' = u'_x + v'_y$, that is , $\zeta = \delta'$ and the latter may be evaluated using $A^{t\prime}$. What may be found from the reflection $(u' = u,\ v' = -v)$, and from the reflection $(u' = v,\ v' = u)$? □

Molinari and Kirwan (1975) use drifter data from the Caribbean Sea to estimate the divergence δ, relative vorticity ζ, stretching deformation $u_x - v_y$ and shearing deformation $v_x + u_y$, using both the least-squares method ($N = 3$, zero residuals) and the area method; see for example Figure 16.3. The agreement is striking, implying that these combinations of spatial velocity gradients have large spatial scales. Four drifters are available for some of the cruise legs ("leg 3") analysed in Figure 16.3. Best fits and residuals are shown in Fig 16.4; the divergences, etc., are much the same as for $N = 3$. The summed residual products $\sum u_n''^2, \sum v_n''^2, \sum u_n'' v_n''$ are relatively smooth for this cruise leg, even though the number of degrees of freedom ($N - 3$) is only one. The correlation between residuals is evidently small, leading Molinari and Kirwin to infer that *"there was little flux of x directed momentum in the y direction, and vice versa, when viewed from a coordinate system moving with the mean velocity"*. Molinari and Kirwan (1975) also diagnose the Cartesian β-plane vorticity budget

$$\frac{\partial}{\partial t}(\zeta + f) + (\zeta + f)\delta = r, \tag{16.20}$$

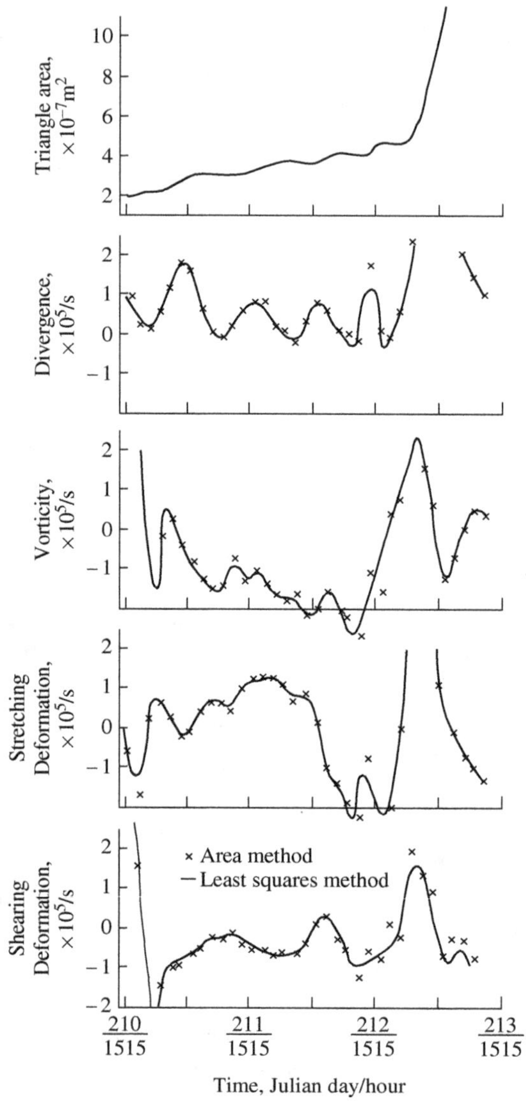

Figure 16.3 The triangle areas *(top panel)* and the differential kinematic properties *(lower panels)* computed for buoy triads, after Molinari and Kirwan (1975).

where the time derivative is Lagrangian, while f is the Coriolis parameter and r is the residual. The terms on the lhs do tend to cancel; see Fig 16.5.

Kirwan (1988) cautions that expansion of the Eulerian velocity field at each drifter position about the same point (the drifter centroid) may be too restrictive. Yet alternatives require the estimation of too many parameters

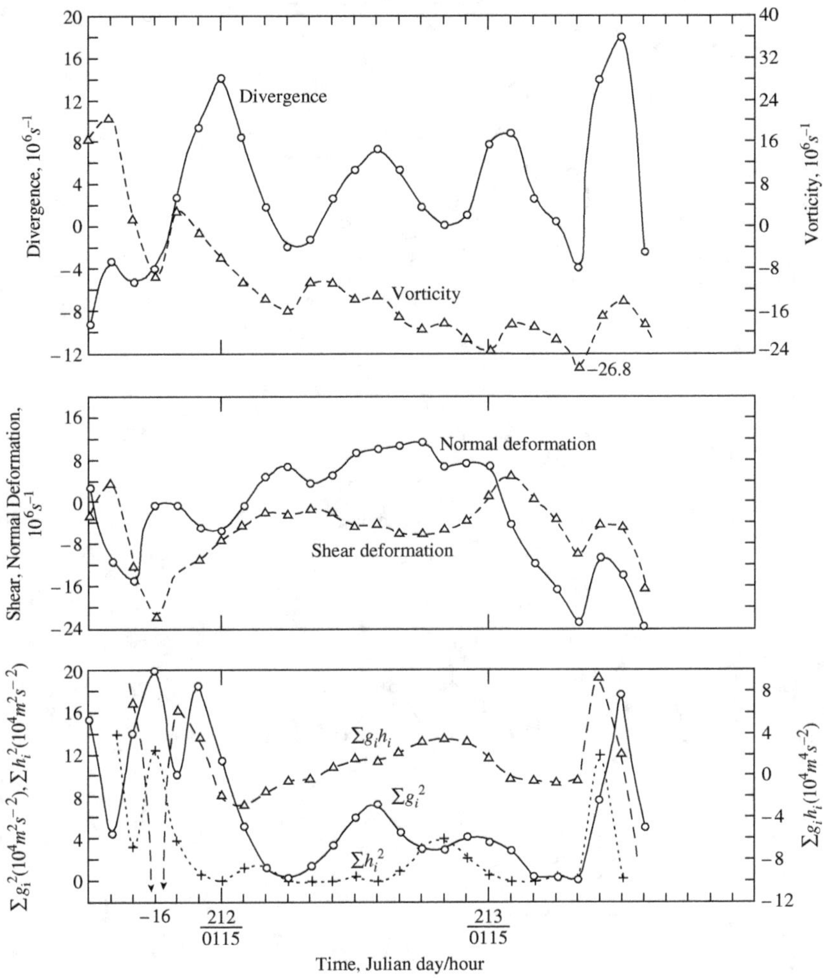

Figure 16.4 Differential kinematic properties and summed residuals (g_i, $h_i = u_i''$, v_i'' here) as calculated from observations of four drifters. The axis scales for the properties exceed those in Figure 16.3 by a factor of 10; after Molinari and Kirwan (1975).

from too little data. Kirwan (1988) also advocates an analysis based on fitting drifter positions, as proposed by Perry and Chong (1987). The estimation problem would be nonlinear: consider the dependence of the solutions of (9.2) upon the velocity gradients. In conclusion, it should be noted that all these methods require at least three drifters, and that they are purely kinematical: no dynamical constraints are imposed on the spatial gradients of velocity.

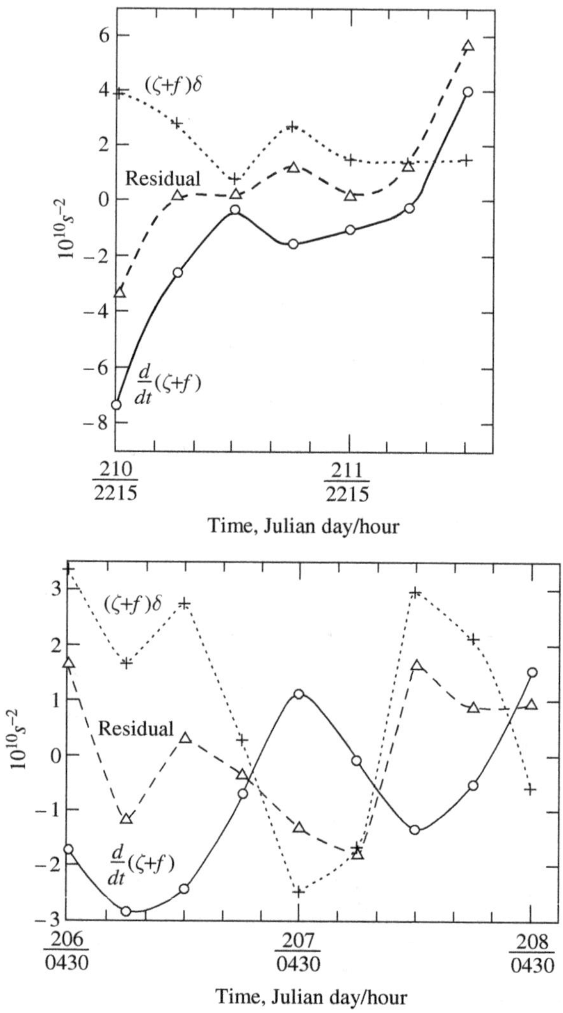

Figure 16.5 Material derivative, stretching and residual terms for the vorticity budget; after Molinari and Kirwan (1975).

16.2.2 Lagrangian dynamical analysis: shallow-water theory

Releases of floats in clusters of three or more are extremely rare, while pairs are more common. There is a need for a spatial gradient analysis, that compensates for the insufficiency of the data by exploiting dynamics. Yet dynamical constraints raise the differential order with respect to time and hence increase the impact of signal noise, so an essential consideration is the availability of first integrals of the dynamics. Finally, fast yet simple optimization algorithms are essential, especially for analysis at sea, as the estimation problem is nonlinear and may be nonsmooth (Ingber, 1993).

The horizontal paths of a pair of drifters or floats or balloons, or their horizontally projected paths, may be expressed in Lagrangian notation as $(X(a, b, s|t), Y(a, b, s|t))$ and $(X(a+\Delta a, b+\Delta b, s+\Delta s|t), Y(a+\Delta a, b+\Delta b, s+\Delta s|t))$. The first-mentioned float is released at position (a, b), at time s; the second at position $(a+\Delta a, b+\Delta b)$, at time $s+\Delta s$. It is assumed that the pair is released on the same pressure surface and remains on that surface (isobaric floats, including drifters), or else on the same density surface and remaining on that surface (isopycnal floats). Assume that the positions of both particles are known for some time interval $s<t<t_1$. If the pair are released sufficiently closely together, it is possible to approximate their subsequent separation as

$$\Delta X(a, b, s|t) = \Delta a \frac{\partial}{\partial a} X(a, b, s|t) + \Delta b \frac{\partial}{\partial b} X(a, b, s|t) + \Delta s \frac{\partial}{\partial s} X(a, b, s|t), \tag{16.21}$$

$$\Delta Y(a, b, s|t) = \Delta a \frac{\partial}{\partial a} Y(a, b, s|t) + \Delta b \frac{\partial}{\partial b} Y(a, b, s|t) + \Delta s \frac{\partial}{\partial s} Y(a, b, s|t). \tag{16.22}$$

Thus, two linear combinations of the six partial derivatives of float position with respect to the release parameters or labels (a, b, s) are known. However, it is desirable to know not just these two combinations of the partials, but in fact all six partials separately. The labeling theorem (1.4) eliminates two of the six:

$$\frac{\partial}{\partial s} X(a, b, s|t) + u[a, b, s] \frac{\partial}{\partial a} X(a, b, s|t) + v[a, b, s] \frac{\partial}{\partial b} X(a, b, s|t) = 0, \tag{16.23}$$

$$\frac{\partial}{\partial s} Y(a, b, s|t) + u[a, b, s] \frac{\partial}{\partial a} Y(a, b, s|t) + v[a, b, s] \frac{\partial}{\partial b} Y(a, b, s|t) = 0. \tag{16.24}$$

The Eulerian velocity of each float is known at release, being the Lagrangian time derivative of the path:

$$\begin{aligned}\big(u[a, b, s], v[a, b, s]\big) &\equiv \big(u(a, b, s|s), v(a, b, s|s)\big) \\ &= \frac{\partial}{\partial t}\big(X(a, b, s|t), Y(a, b, s|t)\big)\Big|_{t=s}.\end{aligned} \tag{16.25}$$

Indeed, if q is any scalar quantity measured by both floats then

$$\Delta q(a,b,s|t) = \Delta a \frac{\partial}{\partial a} q(a,b,s|t) + \Delta b \frac{\partial}{\partial b} q(a,b,s|t) + \Delta s \frac{\partial}{\partial s} q(a,b,s|t) \tag{16.26}$$

is known, subject to the labeling theorem:

$$\frac{\partial}{\partial s} q(a,b,s|t) + u[a,b,s] \frac{\partial}{\partial a} q(a,b,s|t) + v[a,b,s] \frac{\partial}{\partial b} q(a,b,s|t) = 0. \tag{16.27}$$

For example, the pressure p, temperature T and depth h are available for isopycnal floats.

Further constraints on the partials for position and depth of a float may be constructed from the dynamics of the fluid. Assume that the float position and depth satisfy shallow-water equations on the Cartesian β plane:

$$\frac{\partial^2 X}{\partial t^2} - f \frac{\partial Y}{\partial t} = -\frac{g'}{J_s^t} \frac{\partial(h,Y)}{\partial(a,b)}, \tag{16.28}$$

$$\frac{\partial^2 Y}{\partial t^2} + f \frac{\partial X}{\partial t} = -\frac{g'}{J_s^t} \frac{\partial(X,h)}{\partial(a,b)}, \tag{16.29}$$

where $f = f_0 + \beta(Y - y_0)$ is the latitude-dependent Coriolis parameter, with constant f_0, β and y_0, while g' is a reduced value of the gravitational constant and J_s^t is the Jacobi determinant of the transformation $(a,b) \to (X,Y)$. An alternative Lagrangian expression of shallow-water dynamics is obtained by straining the dynamics, that is, by multiplying (16.28), (16.29) with the Jacobi matrix:

$$\frac{\partial X}{\partial a}\left(\frac{\partial^2 X}{\partial t^2} - f\frac{\partial Y}{\partial t}\right) + \frac{\partial Y}{\partial a}\left(\frac{\partial^2 Y}{\partial t^2} + f\frac{\partial X}{\partial t}\right) = -g'\frac{\partial h}{\partial a}, \tag{16.30}$$

$$\frac{\partial X}{\partial b}\left(\frac{\partial^2 X}{\partial t^2} - f\frac{\partial Y}{\partial t}\right) + \frac{\partial Y}{\partial b}\left(\frac{\partial^2 Y}{\partial t^2} + f\frac{\partial X}{\partial t}\right) = -g'\frac{\partial h}{\partial b}. \tag{16.31}$$

These equations may be integrated once with respect to time t, yielding

$$\frac{\partial X}{\partial a}\dot{X} + \frac{\partial Y}{\partial a}\frac{\partial Y}{\partial t} + \frac{\partial \varphi}{\partial a} = U, \tag{16.32}$$

$$\frac{\partial X}{\partial b}\dot{X} + \frac{\partial Y}{\partial b}\frac{\partial Y}{\partial t} + \frac{\partial \varphi}{\partial b} = V, \tag{16.33}$$

where U and V are related to the initial velocity components:

$$U = U[a,b,s] \equiv u(a,b,s|s) - f_1 b - \frac{1}{2}\beta b^2, \tag{16.34}$$

$$V = V[a,b,s] \equiv v(a,b,s|s), \tag{16.35}$$

$f_1 \equiv f_0 - \beta y_0$ being a background rotation, while

$$\dot{X} \equiv \frac{\partial X}{\partial t} - f_1 Y - \frac{1}{2}\beta Y^2 , \tag{16.36}$$

and φ is the Cauchy–Weber integral scalar.

Exercise 16.4 Show that the Cauchy–Weber integral scalar φ is

$$\varphi = \int_s^t \left(g'h - \frac{1}{2}\left(\frac{\partial X}{\partial t}\right)^2 - \frac{1}{2}\left(\frac{\partial Y}{\partial t}\right)^2 + f_1 Y \frac{\partial X}{\partial t} + \frac{1}{2}\beta Y^2 \frac{\partial X}{\partial t} \right) ds' . \tag{16.37}$$

□

Notes

(i) The float velocity (u, v) at the release time $t = s$, and hence the integration constant (U, V), is known from the float position (X, Y) as a function of time t. However (U, V) is independent of time t.

(ii) Integrating by parts the term in (16.37) proportional to f_1 reveals the skew-symmetry of the effect of the background rotation.

(iii) The entire integrand in (16.37) may be evaluated from the position (X, Y) and depth h of a single float, and is therefore a scalar which may be regarded as having been observed by the float. For a pair of floats, the difference in this scalar is

$$\Delta\varphi(a, b, s|t) = \Delta a \frac{\partial}{\partial a}\varphi(a, b, s|t) + \Delta b \frac{\partial}{\partial b}\varphi(a, b, s|t) + \Delta s \frac{\partial}{\partial s}\varphi(a, b, s|t) , \tag{16.38}$$

while the labeling theorem implies

$$\frac{\partial}{\partial s}\varphi(a, b, s|t) + u[a, b, s]\frac{\partial}{\partial a}\varphi(a, b, s|t) + v[a, b, s]\frac{\partial}{\partial b}\varphi(a, b, s|t) = 0 . \tag{16.39}$$

In summary, at any time t the observed values of $(\Delta X, \Delta Y, \Delta\varphi, U, V)$ for a pair of initially close floats yield five linear equations (16.21), (16.22), (16.38), (16.32), (16.33) for the nine partials of (X, Y, φ) with respect to (a, b, s), the coefficients in the linear equations being determined by the observed values of (X, Y). The labeling theorem provides three homogeneous linear equations (16.23), (16.24), (16.39) for the nine partials, with coefficients also being determined by (X, Y). The rank of the eight linear constraint on the nine unknowns is seven. Indeed, the constraints imply that, and can only be solved if the observations $(\Delta X, \Delta Y, \Delta\varphi, U, V)$ satisfy

$$\Delta X\dot{X} + \Delta Y\frac{\partial Y}{\partial t} + \Delta\varphi - U\left(\Delta a - \Delta s(U + f_1 b + \frac{1}{2}\beta b^2)\right) - V\big(\Delta b - \Delta s V\big) = 0\,. \tag{16.40}$$

The solubility condition expresses the orthogonality of the observation vector and the single left null vector of the 8×9 rank-deficient coefficient matrix. Note that the applications of the labeling theorem introduce three trivial or zero-valued observations, for a total of eight observations. If (16.40) is satisfied, then the solution for the nine partials is an indeterminate linear combination of the five nontrivial observations, with two degrees of freedom. Of the five observations, three $(\Delta X, \Delta Y, \Delta\varphi)$ are functions of time t, while two (U, V) are independent of t. If (16.40) is not satisfied, then there is a solution for the nine partials which has the same linear form, but it does not exactly meet the eight constraints. The Moore–Penrose inverse or generalized inverse of the rank deficient matrix of constraints yields the solution with the least Euclidean norm (in nine dimensions), the associated residual having the least Euclidean norm (in eight dimensions). In particular, the Moore–Penrose solution for the observations plus residuals satisfies the solubility condition. See, for example, Lanczos (1966), Bennett (1992) or Wunsch (1996).

The partials of the float positions are subject to two more constraints: the Lagrangian forms of the conservation equations for mass and the Cauchy invariant. Both equations are quadratic in the partials, with coefficients being determined by the float positions and depths. These two nonlinear constraints may further restrict the two degrees of freedom left in the nine partials by the eight linear constraints of rank seven, although not necessarily in an independent or even unique way. Before stating the combined nonlinear least-squares problem for the partials, it is convenient to make a simplifying assumption of little practical restriction, namely, that $\Delta s = 0$. In other words, the two floats are released simultaneously at time $t = s$, or more generally are first both observed at time $t = s$. It follows that the partials with respect to s do not contribute to $(\Delta X, \Delta Y, \Delta\varphi)$, and so the three instances of the labeling theorem do not constrain the partials with respect to (a, b). The labeling theorem may be used to infer the partials with respect to s, once those with respect to (a, b) have been estimated.

Admitting residuals $r_X = r_X(a, b, s|t)$, etc., in all constraints, the time-dependent observations are

$$\Delta a \frac{\partial}{\partial a} X(a, b, s|t) + \Delta b \frac{\partial}{\partial b} X(a, b, s|t) = \Delta X + r_X(a, b, s|t), \tag{16.41}$$

$$\Delta a \frac{\partial}{\partial a} Y(a, b, s|t) + \Delta b \frac{\partial}{\partial b} Y(a, b, s|t) = \Delta Y + r_Y(a, b, s|t), \tag{16.42}$$

$$\Delta a \frac{\partial}{\partial a} \varphi(a, b, s|t) + \Delta b \frac{\partial}{\partial b} \varphi(a, b, s|t) = \Delta \varphi + r_\varphi(a, b, s|t). \tag{16.43}$$

The time-independent observations (the Cauchy–Weber integration constants U, V) appear in the dynamical constraints

$$\left(\frac{\partial}{\partial a} X(a, b, s|t)\right) \dot{X}(a, b, s|t) + \left(\frac{\partial}{\partial a} Y(a, b, s|t)\right) \frac{\partial}{\partial t} Y(a, b, s|t) + \frac{\partial}{\partial a} \varphi(a, b, s|t) = U[a, b, s] + r_U(a, b, s|t), \tag{16.44}$$

$$\left(\frac{\partial}{\partial b} X(a, b, s|t)\right) \dot{X}(a, b, s|t) + \left(\frac{\partial}{\partial b} Y(a, b, s|t)\right) \frac{\partial}{\partial t} Y(a, b, s|t) + \frac{\partial}{\partial b} \varphi(a, b, s|t) = V(a, b, s) + r_V(a, b, s|t). \tag{16.45}$$

Notes

(i) The six state variables are the partials of X, Y and φ with respect to a and b.

(ii) The five observed quantities are $(\Delta X, \Delta Y, \Delta\varphi, U, V)$.

(iii) The five linear constraints (16.41)–(16.45) are described as time dependent or not, according to the nature of the observations. The coefficients in the two dynamical constraints are time dependent, and are known from the float data. The rank of the five constraints is four.

(iv) The constraints are now *weak*, owing to the admission of five nonvanishing residuals $r_X, r_Y, r_\varphi, r_U, r_V$. All residuals are time dependent.

(v) The solubility condition for (16.41)–(16.45) is

$$(\Delta X + r_X)\dot{X} + (\Delta Y + r_Y)\frac{\partial Y}{\partial t} + \Delta\varphi + r_\varphi - \Delta a\,(U + r_U) - \Delta b\,(V + r_V) = \Xi \tag{16.46}$$

where $\Xi = \Xi(t)$ is identically zero:

$$\Xi \equiv 0. \tag{16.47}$$

Note that (16.46) and (16.47) constitute a linear constraint upon the residuals $(r_X, r_Y, r_\varphi, r_U, r_V)$.

After one integration in time, the Lagrangian form of the law of volume conservation, or continuity equation, for shallow-water dynamics is

$$h(a,b,s|t)\left(\left(\frac{\partial}{\partial a}X(a,b,s|t)\right)\frac{\partial}{\partial b}Y(a,b,s|t)-\left(\frac{\partial}{\partial b}X(a,b,s|t)\right)\frac{\partial}{\partial a}Y(a,b,s|t)\right)$$
$$= h[a,b,s]+r_h(a,b,s|t)\,. \tag{16.48}$$

This sixth equation is quadratic in four of the six partials. There is a sixth observed quantity $h[a, b, s]$, and a sixth residual $r_h(a, b, s|t)$. The time-dependent coefficient $h(a, b, s|t)$ and time-independent observation $h[a, b, s]$ are known float depths. The Lagrangian form for the law of conservation of the Cauchy invariant is obtained by cross-differentation of the Cauchy–Weber integrated momentum equations (16.44) and (16.45), yielding

$$\left(\frac{\partial}{\partial b}X(a,b,s|t)\right)\frac{\partial^2}{\partial a\partial t}X(a,b,s|t)+\left(\frac{\partial}{\partial b}Y(a,b,s|t)\right)\frac{\partial^2}{\partial a\partial t}Y(a,b,s|t)$$
$$-\left(\frac{\partial}{\partial a}X(a,b,s|t)\right)\frac{\partial^2}{\partial b\partial t}X(a,b,s|t)-\left(\frac{\partial}{\partial a}Y(a,b,s|t)\right)\frac{\partial^2}{\partial b\partial t}Y(a,b,s|t)$$
$$+f\big(Y(a,b,s|t)\big)J(a,b,s|t)=\zeta[a,b,s]+r_\zeta(a,b,s|t)\,. \tag{16.49}$$

Notes

(i) The seventh equation (16.49) is an nonlinear ordinary differential equation in time t for four of the six partials. There is a seventh observed quantity $\zeta[a, b, s]$, and a seventh residual $r_\zeta(a, b, s|t)$.
(ii) The Coriolis parameter $f(Y)$ is known on the float path.
(iii) $J(a, b, s|t)$ or J_s^t is the Jacobi determinant of the float cooordinate transformation $(a, b) \to (X, Y)$. It appears also in the continuity equation (16.46), since

$$J(a,b,s|t)\equiv\frac{\partial(X,Y)}{\partial(a,b)}(a,b,s|t)$$
$$\equiv\left(\frac{\partial}{\partial a}X(a,b,s|t)\right)\frac{\partial}{\partial b}Y(a,b,s|t)$$
$$-\left(\frac{\partial}{\partial b}X(a,b,s|t)\right)\frac{\partial}{\partial a}Y(a,b,s|t)\,. \tag{16.50}$$

(iv) The total vorticity at release

$$\zeta[a,b,s] \equiv \frac{\partial}{\partial a}V[a,b,s] - \frac{\partial}{\partial b}U[a,b,s] + f(b) \tag{16.51}$$

is not known from the float path. If the planetary vorticity at release $f(Y) = f(b)$ serves a default estimate then $r_\zeta(a,b,s|s)$, the ζ residual at release, is the unknown relative vorticity at release.

(v) The nonlinearity of the conservation law for volume and for the Cauchy invariant preclude direct use of the Moore–Penrose inverse. Before exploring alternative optimization algorithms, it is appropriate to consider more realistic dynamics.

16.2.3 Lagrangian dynamical analysis: Boussinesq theory

Shallow-water theory serves for a preliminary exposition of Lagrangian dynamical analysis. The minimal resolution of such a $1\frac{1}{2}$ layer model is inadequate for the analysis of real float data, as the depth and velocity of an isopycnal float cannot be accurately identified with the shallow-water state variables. Moreover, the great majority of floats have constant mass and volume, and settle to a depth of the same *in situ* density. These are nominally "isobaric", since the mean density profile is approximately a function of pressure alone. Such floats do record changes in pressure, which in a first approximation may be attributed to changing stratification owing either to local changes or to the horizontal motion of the float.[1] The inescapable inference is that float pair analysis must be constrained by continuously stratified Lagrangian fluid dynamics. Sound waves play no dynamical role in ocean circulation,[2] and anelastic effects are most likely negligible given the precision of the data and the estimation method, so a Boussinesq fluid is assumed (Phillips, 1966). Spherical polar coordinates are needed for the analysis of float tracks of many months' duration, but the β-plane approximation clarifies a first discussion and is likely adequate for shorter tracks. The verbose notation (X, Y, Z) for the three-dimensional (eastward, northward, vertical) position of a float at time t, with initial positions (a, b, c) at the release time s, will be clearer than subscripted notation owing to the asymmetries of the Coriolis and buoyant accelerations. Conservation of momentum takes the forms

$$\frac{\partial^2 X}{\partial t^2} - f\frac{\partial Y}{\partial t} = -\frac{1}{\rho_0 J_s^t}\frac{\partial(p^{\text{dyn}}, Y, Z)}{\partial(a,b,c)}, \tag{16.52}$$

[1] For a thorough investigation of float dynamics, see Goodman and Levine (1990).
[2] Aside from float tracking!

$$\frac{\partial^2 Y}{\partial t^2}+f\frac{\partial X}{\partial t}=-\frac{1}{\rho_0 J_s^t}\frac{\partial(X,p^{\rm dyn},Z)}{\partial(a,b,c)}\,, \tag{16.53}$$

$$\frac{\partial^2 Z}{\partial t^2}=-\frac{1}{\rho_0 J_s^t}\frac{\partial(X,Y,p^{\rm dyn})}{\partial(a,b,c)}-g\frac{\rho^{\rm dyn}}{\rho_0}\,, \tag{16.54}$$

where $f=f_0+\beta(Y-y_0)$, and where $p^{\rm dyn}$ and $\rho^{\rm dyn}$ are the dynamical departures from static profiles $p^{\rm stat}$ and $\rho^{\rm stat}$. That is,

$$p(a,b,c,s|t)=p^{\rm stat}(c)+p^{\rm dyn}(a,b,c,s|t)\,, \tag{16.55}$$

$$\rho(a,b,c,s|t)=\rho^{\rm stat}(c)+\rho^{\rm dyn}(a,b,c,s|t)\,, \tag{16.56}$$

satisfying

$$\frac{d}{dc}p^{\rm stat}(c)=-g\rho^{\rm stat}(c)\,, \tag{16.57}$$

g being the gravitational acceleration. Note that according to (16.54) the dynamic departures $p^{\rm dyn}$ and $\rho^{\rm dyn}$ are not assumed to be in hydrostatic balance. In the limit of low Mach number, mass is approximately conserved by the flow if

$$J_s^t\equiv\frac{\partial(X,Y,Z)}{\partial(a,b,c)}=J_s^s=1\,, \tag{16.58}$$

and internal energy is approximately conserved if

$$\frac{\partial\rho}{\partial t}=\frac{\partial\rho^{\rm dyn}}{\partial t}=0\,. \tag{16.59}$$

Multiplying the momentum conservation equations by the Jacobi matrix yields

$$\frac{\partial X}{\partial a}\left(\frac{\partial^2 X}{\partial t^2}-f\frac{\partial Y}{\partial t}\right)+\frac{\partial Y}{\partial a}\left(\frac{\partial^2 Y}{\partial t^2}+f\frac{\partial X}{\partial t}\right)+\frac{\partial Z}{\partial a}\frac{\partial^2 Z}{\partial t^2}=-\frac{1}{\rho_0}\frac{\partial p^{\rm dyn}}{\partial a}-g\frac{\partial Z}{\partial a}\frac{\rho^{\rm dyn}}{\rho_0}\,, \tag{16.60}$$

$$\frac{\partial X}{\partial b}\left(\frac{\partial^2 X}{\partial t^2}-f\frac{\partial Y}{\partial t}\right)+\frac{\partial Y}{\partial b}\left(\frac{\partial^2 Y}{\partial t^2}+f\frac{\partial X}{\partial t}\right)+\frac{\partial Z}{\partial b}\frac{\partial^2 Z}{\partial t^2}=-\frac{1}{\rho_0}\frac{\partial p^{\rm dyn}}{\partial b}-g\frac{\partial Z}{\partial b}\frac{\rho^{\rm dyn}}{\rho_0}\,, \tag{16.61}$$

$$\frac{\partial X}{\partial c}\left(\frac{\partial^2 X}{\partial t^2}-f\frac{\partial Y}{\partial t}\right)+\frac{\partial Y}{\partial c}\left(\frac{\partial^2 Y}{\partial t^2}+f\frac{\partial X}{\partial t}\right)+\frac{\partial Z}{\partial c}\frac{\partial^2 Z}{\partial t^2}=-\frac{1}{\rho_0}\frac{\partial p^{\rm dyn}}{\partial c}-g\frac{\partial Z}{\partial c}\frac{\rho^{\rm dyn}}{\rho_0}\,. \tag{16.62}$$

After one integration with respect to time, (16.60), (16.61) and (16.62) become

$$\frac{\partial X}{\partial a}\dot{X}+\frac{\partial Y}{\partial a}\frac{\partial Y}{\partial t}+\frac{\partial Z}{\partial a}\frac{\partial Z}{\partial t}+\frac{\partial \varphi}{\partial a}=U+\frac{g}{\rho_0}\int_s^t Zdt'\frac{\partial \rho^{\mathrm{dyn}}}{\partial a}, \tag{16.63}$$

$$\frac{\partial X}{\partial b}\dot{X}+\frac{\partial Y}{\partial b}\frac{\partial Y}{\partial t}+\frac{\partial Z}{\partial b}\frac{\partial Z}{\partial t}+\frac{\partial \varphi}{\partial b}=V+\frac{g}{\rho_0}\int_s^t Zdt'\frac{\partial \rho^{\mathrm{dyn}}}{\partial b}, \tag{16.64}$$

$$\frac{\partial X}{\partial c}\dot{X}+\frac{\partial Y}{\partial c}\frac{\partial Y}{\partial t}+\frac{\partial Z}{\partial c}\frac{\partial Z}{\partial t}+\frac{\partial \varphi}{\partial c}=W+\frac{g}{\rho_0}\int_s^t Zdt'\frac{\partial \rho^{\mathrm{dyn}}}{\partial c}, \tag{16.65}$$

where

$$U=U[a,\,b,\,c,\,s]\equiv u(a,\,b,\,c,\,s|s)-f_1 b-\frac{1}{2}\beta b^2\,, \tag{16.66}$$

$$V=V[a,\,b,\,c,\,s]\equiv v(a,\,b,\,c,\,s|s)\,, \tag{16.67}$$

$$W=W[a,\,b,\,c,\,s]\equiv w(a,\,b,\,c,\,s|s)\,, \tag{16.68}$$

while $\dot{X}$ is given by (16.36), and φ is the Cauchy–Weber integral scalar.

Exercise 16.5 Show that the Cauchy–Weber integral scalar φ is

$$\varphi=\int_s^t\left(\frac{p^{\mathrm{dyn}}}{\rho_0}+\frac{\rho^{\mathrm{dyn}}}{\rho_0}gZ-\frac{1}{2}\left(\frac{\partial X}{\partial t}\right)^2-\frac{1}{2}\left(\frac{\partial Y}{\partial t}\right)^2 -\frac{1}{2}\left(\frac{\partial Z}{\partial t}\right)^2+f_1 Y\frac{\partial X}{\partial t}+\frac{1}{2}\beta Y^2\frac{\partial X}{\partial t}\right)ds'\,. \tag{16.69}$$

□

The Cauchy vector is not conserved, owing to the action of buoyancy:

$$\varpi_1+f\frac{\partial(X,\,Y)}{\partial(b,\,c)}=\frac{\partial W}{\partial b}-\frac{\partial V}{\partial c}-\frac{g}{\rho_0}\int_s^t\left(\frac{\partial Z}{\partial b}\frac{\partial \rho^{\mathrm{dyn}}}{\partial c}-\frac{\partial Z}{\partial c}\frac{\partial \rho^{\mathrm{dyn}}}{\partial b}\right)dt'\,, \tag{16.70}$$

where

$$\varpi_1\equiv\frac{\partial Y}{\partial c}\frac{\partial^2 Y}{\partial b\partial t}+\frac{\partial Z}{\partial c}\frac{\partial^2 Z}{\partial b\partial t}+\frac{\partial X}{\partial c}\frac{\partial^2 X}{\partial b\partial t}-\frac{\partial Y}{\partial b}\frac{\partial^2 Y}{\partial c\partial t}-\frac{\partial Z}{\partial b}\frac{\partial^2 Z}{\partial c\partial t}-\frac{\partial X}{\partial b}\frac{\partial^2 X}{\partial c\partial t}; \tag{16.71}$$

$$\varpi_2+f\frac{\partial(X,\,Y)}{\partial(c,\,a)}=\frac{\partial U}{\partial c}-\frac{\partial W}{\partial a}-\frac{g}{\rho_0}\int_s^t\left(\frac{\partial Z}{\partial c}\frac{\partial \rho^{\mathrm{dyn}}}{\partial a}-\frac{\partial Z}{\partial a}\frac{\partial \rho^{\mathrm{dyn}}}{\partial c}\right)dt'\,, \tag{16.72}$$

where

$$\varpi_2 \equiv \frac{\partial Z}{\partial a}\frac{\partial^2 Z}{\partial c \partial t} + \frac{\partial X}{\partial a}\frac{\partial^2 X}{\partial c \partial t} + \frac{\partial Y}{\partial a}\frac{\partial^2 Y}{\partial c \partial t} - \frac{\partial Z}{\partial c}\frac{\partial^2 Z}{\partial a \partial t} - \frac{\partial X}{\partial c}\frac{\partial^2 X}{\partial a \partial t} - \frac{\partial Y}{\partial c}\frac{\partial^2 Y}{\partial a \partial t}; \tag{16.73}$$

and

$$\varpi_3 + f\frac{\partial(X, Y)}{\partial(a, b)} = \frac{\partial V}{\partial a} - \frac{\partial U}{\partial b} - \frac{g}{\rho_0}\int_s^t \left(\frac{\partial Z}{\partial a}\frac{\partial \rho^{\mathrm{dyn}}}{\partial b} - \frac{\partial Z}{\partial b}\frac{\partial \rho^{\mathrm{dyn}}}{\partial a}\right) dt', \tag{16.74}$$

where

$$\varpi_3 \equiv \frac{\partial X}{\partial b}\frac{\partial^2 X}{\partial a \partial t} + \frac{\partial Y}{\partial b}\frac{\partial^2 Y}{\partial a \partial t} + \frac{\partial Z}{\partial b}\frac{\partial^2 Z}{\partial a \partial t} - \frac{\partial X}{\partial a}\frac{\partial^2 X}{\partial b \partial t} - \frac{\partial Y}{\partial a}\frac{\partial^2 Y}{\partial b \partial t} - \frac{\partial Z}{\partial a}\frac{\partial^2 Z}{\partial b \partial t}. \tag{16.75}$$

Buoyancy will vary the Cauchy vector ϖ_i unless the dynamic density ρ^{dyn} is a function of float depth Z alone. Owing to (16.59), such circumstances are possible only if ρ^{dyn} is a constant. Note the disposition of ρ^{dyn} inside the time integral, in anticipation of a weak reformulation of the Boussinesq dynamics admitting time-dependent residuals in the energy equation (16.59).

The pair differences for the circulation or state variables $(X, Y, Z, \varphi, \rho^{\mathrm{dyn}})$ are

$$\Delta X = \Delta a\frac{\partial X}{\partial a} + \Delta b\frac{\partial X}{\partial b} + \Delta c\frac{\partial X}{\partial c}, \tag{16.76}$$

$$\Delta Y = \Delta a\frac{\partial Y}{\partial a} + \Delta b\frac{\partial Y}{\partial b} + \Delta c\frac{\partial Y}{\partial c}, \tag{16.77}$$

$$\Delta Z = \Delta a\frac{\partial Z}{\partial a} + \Delta b\frac{\partial Z}{\partial b} + \Delta c\frac{\partial Z}{\partial c}, \tag{16.78}$$

$$\Delta \varphi = \Delta a\frac{\partial \varphi}{\partial a} + \Delta b\frac{\partial \varphi}{\partial b} + \Delta c\frac{\partial \varphi}{\partial c}, \tag{16.79}$$

$$\Delta \rho^{\mathrm{dyn}} = \Delta a\frac{\partial \rho^{\mathrm{dyn}}}{\partial a} + \Delta b\frac{\partial \rho^{\mathrm{dyn}}}{\partial b} + \Delta c\frac{\partial \rho^{\mathrm{dyn}}}{\partial c}, \tag{16.80}$$

to first order in Δa, etc. Note that simultaneous releases are assumed: $\Delta s = 0$.

There are 15 Lagrangian spatial partial derivatives of $(X, Y, Z, \varphi, \rho^{\mathrm{dyn}})$ with respect to (a, b, c), to be determined from the five observations (16.76)–(16.80). These linear constraints are complemented by the three linear constraints (16.63)–(16.65) containing the observed Cauchy–Weber integration constants (U, V, W), by the three nonlinear constraints (16.70)–(16.75) for the Cauchy vector and the initial total vorticity, and by the one nonlinear constraint (16.58) expressing approximate conservation of mass. To these

12 constraints are added three linear constraints obtained by differentiating the time integral of the energy equation (16.59)

$$\frac{\partial}{\partial a}\rho^{\mathrm{dyn}}(a,b,c,s|t)=\frac{\partial}{\partial a}\rho^{\mathrm{dyn}}(a,b,c,s|s)\,, \tag{16.81}$$

$$\frac{\partial}{\partial b}\rho^{\mathrm{dyn}}(a,b,c,s|t)=\frac{\partial}{\partial b}\rho^{\mathrm{dyn}}(a,b,c,s|s)\,, \tag{16.82}$$

$$\frac{\partial}{\partial c}\rho^{\mathrm{dyn}}(a,b,c,s|t)=\frac{\partial}{\partial c}\rho^{\mathrm{dyn}}(a,b,c,s|s)\,. \tag{16.83}$$

The initial values of the density gradients are known only in a single linear combination but will, like the initial vorticity, be regarded as incompletely known, time-independent observations. The 8 linear constraints (16.76)–(16.80), (16.63)–(16.65) are of rank 7, with solubility condition

$$\Delta X\dot{X}+\Delta Y\frac{\partial Y}{\partial t}+\Delta Z\frac{\partial Z}{\partial t}\Delta\varphi-(\Delta aU+\Delta bV+\Delta cW)-\frac{g}{\rho_0}\int_s^t Z\,dt'\,\Delta\rho^{\mathrm{dyn}}=0\,. \tag{16.84}$$

As with the shallow-water formulation, a least-squares resolution of the problem of Boussinesq Lagrangian dynamical analysis is available, by introducing 15 residuals.

16.2.4 Least-squares estimator

The solutions for the partials of $(X, Y, \dots)$ with respect to $(a, b, \dots)$ are those which minimize a weighted sum of squares of the N residuals $(r_1, \dots, r_N) \equiv (r_X, r_Y, \dots)$ subject to a solubility condition $\Xi = 0$, where Ξ is a linear combination of the observations and residuals. The estimator is

$$\mathcal{J}=\int_s^{t_1}\sum_{i=1}^{N}W_i\left(r_i^2+\tau^2\left(\frac{\partial r_i}{\partial t}\right)^2\right)dt+\int_s^{t_1}\lambda\Xi\,dt\,. \tag{16.85}$$

Notes

(i) The residuals are being penalized not only for magnitude, but also for temporal variability on scales of $O(\tau)$ or smaller.
(ii) The weights W_i will be identified with the reciprocals of the variances of the residuals r_i.
(iii) The strong constraint $\Xi=0$ is being appended to the residual penalty with the Lagrange multiplier $\lambda=\lambda(t)$.
(iv) The penalty $\mathcal{J}$ depends implicitly upon the partials, through the definitions of the residuals. The dependence upon λ and Ξ is explicit.

Exercise 16.6 Let $r(t)$ be any function defined for $-\infty < t < \infty$. Consider the infinite integral

$$\mathcal{J} = \frac{1}{2} W_0 \int_{-\infty}^{\infty} \left(r(t)^2 + \tau^2 \left(\frac{d}{dt} r(t) \right)^2 \right) dt . \tag{16.86}$$

Show that

$$\mathcal{J} = \int_{-\infty}^{\infty} \int_{-\infty}^{\infty} r(t) W(t - t') r(t') dt dt' , \tag{16.87}$$

where

$$W(t) = \frac{1}{2} W_0 \left(1 - \tau^2 \left(\frac{d}{dt} \right)^2 \right) \delta(t) . \tag{16.88}$$

Show further that

$$\int_{-\infty}^{\infty} W(t_1 - t_2) C(t_2 - t_3) dt_2 = \delta(t_1 - t_3) , \tag{16.89}$$

where

$$C(t) = (W_0 \tau)^{-1} \exp\left(-|t|/\tau\right) . \tag{16.90}$$

That is, $\mathcal{J}$ is the estimator of maximum likelihood, for the stationary normal random process $r(t)$ having vanishing mean and covariance $C(t)$. In particular, the variance is

$$C(0) = (W_0 \tau)^{-1} . \tag{16.91}$$

□

The motivation for establishing a relationship between a covariance and a weighting kernel is that it is more natural to hypothesize the covariance of the residuals in an inverse problem, yet miminization of the estimator by searching in state space requires that the weighting kernel be known. Determining the weighting kernel W by explicit inversion of the covariance C, as implied by (16.89), is usually prohibitive so the relationship between the simple Markovian covariance (16.90) and the easily manipulated estimator (16.86) is particularly convenient.

Exercise 16.7 Derive the Euler–Lagrange equations, for extrema of the penalty $\mathcal{J}$ defined by (16.85), with respect to variations of the partials and the Lagrange multiplier. Show that these equations may be reduced to coupled differential equations for the variables adjoint to the continuity and vorticity equations, plus a single differential equation for the Cauchy–Weber observational residual r_φ. This exercise is not trivial. A first attempt should be based on the shallow-water equations. □

References

Abbott, M. R., K. H. Brink, C. R. Booth, D. Blasco, M. S. Swenson, C. O. Davis and L. A. Codispoti, 1995: Scales of variability of bio-optical properties as observed from near-surface drifters. *Journal of Geophysical Research*, **100**, 13345–13367.

Abbott, M. R. and R. M. Letelier, 1998: Decorrelation scales of chlorophyll as observed from bio–optical drifters in the California Current. *Deep-Sea Research II*, **45**, 1639–1667.

Abrashkin, A. A. and E. I. Yakobovich, 1984: Planar rotational flows of an ideal fluid. *Soviet Physics Doklady*, **29** (5), 370–371.

Adams, R. A., 1975: *Sobolev Spaces*. New York, Academic Press.

Anderson, D. L. T., J. Sheinbaum and K. Haines, 1996: Data assimilation in ocean models. *Reports on Progress in Physics*, **59**, 1209–1266.

Andrews, D. G. and M. E. McIntyre, 1978: An exact theory of nonlinear waves on a Lagrangian mean flow. *Journal of Fluid Mechanics*, **89**, 609–646.

Apostol, T. M., 1957: *Mathematical Analysis: A Modern Approach to Advanced Calculus*. Addison–Wesley, Reading.

Baer, F., 1972: An alternate scale representation of atmospheric energy spectra. *Journal of the Atmospheric Sciences*, **29**, 649–664.

Baker, D. J., 1966: A technique for the precise measurement of small fluid velocities. *Journal of Fluid Mechanics*, **29**, 165–175.

Baker, D. J. and A. R. Robinson, 1969: A laboratory model for the general ocean circulation. *Philosophical Transactions of the Royal Society*, **A–265**, 1168, 533–566.

Batchelor, G. K., 1952: Diffusion in a field of homogeneous turbulence, II: the relative motion of particles. *Proceedings of the Cambridge Philosophical Society*, **48**, 345–362.

Batchelor, G. K., 1959: Small-scale variations of convected quantities like temperature in a turbulent fluid, Part 1, General discussion and the case of small conductivity. *Journal of Fluid Mechanics*, **5**, 113–133.

Batchelor, G. K., 1960: *The Theory of Homogeneous Turbulence*. Cambridge, Cambridge University Press.

Batchelor, G. K., 1973: *An Introduction to Fluid Mechanics*. Cambridge, Cambridge University Press.

Batchelor, G. K., I. D. Howells and A. A. Townsend, 1959: Small-scale variations of convected quantities like temperature in turbulent fluid, part 2: The case of large conductivity. *Journal of Fluid Mechanics*, **5**, 134–139.

Bateman, H., 1929: Notes on a differential equation which occurs in the two-dimensional motion of a compressible fluid and the associated variational problem. *Proceedings of the Royal Society of London, Series A*, **125**, 598–618.

Bateman, H., 1944: *Partial Differential Equations of Mathematical Physics*. New York, Dover.

Beardsley, R. C., 1969: A laboratory model of the wind-driven ocean circulation. *Journal of Fluid Mechanics*, **38**, 255–271.

Bennett, A. F. and J. F. Middleton, 1983: Statistical mechanics of a finite-difference approximation to the barotropic vorticity equation. *Quarterly Journal of the Royal Meteorological Society*, **109**, 795–808.

Bennett, A. F., 1984: Relative dispersion: local and nonlocal dynamics. *Journal of the Atmospheric Sciences*, **41**, 1881–1886.

Bennett, A. F., 1986a: High wavenumber spectrum of a passive scalar in isotropic turbulence. *Physics of Fluids*, **29**, 1734–1735.

Bennett, A. F., 1986b: Random walks with manholes: simple models of dispersion in turbulence with coherent vortices. *Journal of Geophysical Research*, **91**, 10769–10770.

Bennett, A. F., 1987: A Lagrangian analysis of turbulent diffusion. *Reviews of Geophysics*, **25**, 799–822.

Bennett, A. F., 1992: *Inverse Methods in Physical Oceanography*. Monographs on Mechanics and Applied Mathematics, Cambridge, Cambridge University Press.

Bennett, A. F., 1996: Particle displacements in inhomogeneous turbulence. In *Stochastic Modeling in Physical Oceanography*, ed. R. J. Adler, P. Müller and B. Rozovskii. Boston, Birkhäuser.

Bennett, A. F., 2002: *Inverse Modeling of the Ocean and Atmosphere*. Cambridge, Cambridge University Press.

Bennett, A. F. and B. S. Chua, 1999: Open boundary conditions for Lagrangian geophysical fluid dynamics. *Journal of Computational Physics*, **153**, 418–436.

Bennett, A. F., B. S. Chua, D. E. Harrison and M. J. McPhaden, 1998: Generalized inversion of Tropical Atmosphere–Ocean (TAO) data and a coupled model of the tropical Pacific. *Journal of Climate*, **11**, 1768–1792.

Bennett, A. F., B. S. Chua, D. E. Harrison and M. J. McPhaden, 2000: Generalized inversion of Tropical Atmosphere–Ocean (TAO) data and a coupled model of the tropical Pacific, II: The 1995–96 La Niña and 1997–98 El Niño. *Journal of Climate*, **13**, 2770–2785.

Bennett, A. F., B. S. Chua, H.-E. Ngodock, D. E. Harrison and M. J. McPhaden, 2005: Generalized inversion of Tropical Atmosphere–Ocean (TAO) data and the Gent–Cane model of the tropical Pacific Ocean. *Journal of Marine Research*, [submitted].

Bennett, A. F. and K. L. Denman, 1985: Phytoplankton patchiness: inferences from particle statistics. *Journal of Marine Research*, **43**, 307–335.

Bennett, A. F. and K. L. Denman, 1989: Large-scale patchiness due to an annual plankton cycle. *Journal of Geophysical Research*, **94**, 823–829.

Bennett, A. F. and P. E. Kloeden, 1981: The dissipative quasigeostrophic equations. *Mathematika*, **28**, 265–285.

Bennett, A. F. and P. C. McIntosh, 1982: Open ocean modeling as an inverse problem: tidal theory. *Journal of Physical Oceanography*, **12**, 1004–1018.

Boer, G. J., and T. G. Shepherd, 1983: Large-scale two dimensional turbulence in the atmosphere. *Journal of the Atmospheric Sciences*, **40**, 164–184.

Bowden, K. F., 1965: Horizontal mixing in the sea due to a shearing current. *Journal of Fluid Mechanics*, **21**, 83–95.

Bretherton, F. P., 1970: A note on Hamilton's principle for perfect fluids. *Journal of Fluid Mechanics*, **44**, 19–31.

Carter, E. F., 1989: Assimilation of Lagrangian data into a model. *Dynamics of Atmospheres and Oceans*, **13**, 335–348.

Champagne, F. H., C. A. Friehe, J. C. LaRue and J. C. Wyngaard, 1977: Flux measurements, flux estimation techniques, and fine-scale turbulence measurements in the unstable surface layer over land. *Journal of the Atmospheric Sciences*, **34**, 515–530.

Chapman, S. and T. G. Cowling, 1970: *The Mathematical Theory of Non-Uniform Gases*, 3rd edn. Cambridge, Cambridge University Press.

Chen, T.-C., and A. Wiin-Nielsen, 1978: Non-linear cascades of atmospheric energy and enstrophy in a two-dimensional spectral index. *Tellus*, **30**, 313–322.

Chen, T. N. and R. A. D. Byron-Scott, 1995: Bifurcation in cross-equatorial airflow: A nonlinear characteristic of Lagrangian modeling. *Journal of the Atmospheric Sciences*, **52**, 1383–1400.

Childress, S., G. R. Ierley, E. A. Spiegel and W. R. Young, 1989: Blow-up of unsteady two-dimensional Euler and Navier-Stokes solutions having stagnation-point form. *Journal of Fluid Mechanics*, **203**, 1–22.

Clay, J. P., 1973: Turbulent mixing of temperature in water, air and mercury. PhD dissertation, Engineering Physics Department, University of California, San Diego.

Cocke, W. J., 1972: Central limit theorem for sums of dependent vector variables. *Annals of Mathematical Statistics*, **43**, 968–976.

Constantin, A., 2001: Edge waves along a sloping beach. *Journal of Physics A*, **34**, 9723–9731.

Corrsin, S., 1951: On the spectrum of isotropic temperature fluctuations in an isotropic turbulence. *Journal of Applied Physics*, **22**, 469–473.

Corrsin, S., 1959: Progress report on some turbulent diffusion research. *Advances in Geophysics*, **6**, 161–164.

Courant, R. and D. Hilbert, 1953: *Methods of Mathematical Physics*, Vol. 1. New York, Wiley Interscience.

Crease, J., 1965: The flow of Norwegian Sea water through the Faroe Bank Channel. *Deep-Sea Research*, **12**, 143–150.

Daley, R. A., 1991: *Atmospheric Data Analysis*. Cambridge, Cambridge University Press.

Danabosoglu, A., J. C. McWilliams and P. R. Gent, 1994: The role of mesoscale tracer transports in the global ocean circulation. *Science*, **264**, 1123–1126.

Davis, R. E., 1985: Drifter observations of coastal surface currents during CODE: The statistical and dynamical views. *Journal of Geophysical Research*, **90**, 4756–4772.

Davis, R. E., 1987: Modeling eddy transports of passive tracers. *Journal of Marine Research*, **45**, 635–666.

Davis, R. E., D. C. Webb, L. A. Regier and J. Dufour, 1992: The Autonomous Lagrangian Circulation Explorer (ALACE). *Journal of Atmospheric and Oceanic Technology*, **9**, 264–285.

Davis, R. E. and W. Zenk, 2001: Subsurface Lagrangian observations during the 1990s. In *Ocean Circulation and Climate: Observing and Modelling the Global Ocean*, ed. G. Siedler, J. Church and J. Gould. San Diego, Academic Press.

Drazin, P. G. and W. H. Reid, 1981: *Hydrodynamic Stability*. Cambridge, Cambridge University Press.

Drummond, I. T., 1982: Path-integral methods for turbulent diffusion. *Journal of Fluid Mechanics*, **123**, 59–68.

Durran, D. R., 1999: *Numerical Methods for Wave Equations in Geophysical Fluid Dynamics*. New York, Springer.

Dvorkin, Y. and N. Paldor, 1999: Analytical considerations of Lagrangian cross-equatorial flow. *Journal of the Atmospheric Sciences*, **56**, 1229–1237.

Eckart, C., 1960: Variational principles of hydrodynamics. *Physics of Fluids*, **3**, 421–427.

Elliott, F. W. and A. Majda, 1996: Pair dispersion over an inertial range spanning many decades. *Physics of Fluids*, **8**, 1052–1060.

Emery, W. J. and R. E. Thomson, 1997: *Data Analysis Methods in Physical Oceanography*. New York, Pergamon.

Erdélyi, A., W. Magnus, F. Oberhettinger and F. G. Tricomi, 1954: *Tables of Integral Transforms*, Vol. 1, New York, McGraw-Hill.

Er-El, J. and R. I. Peskin, 1981: Relative diffusion of constant level balloons in the southern hemisphere. *Journal of the Atmospheric Sciences*, **38**, 2264–2274.

Fahrbach, E., C. Brockmann and J. Meincke, 1986: Horizontal mixing in the Atlantic Equatorial Undercurrent estimated from drifting buoy clusters. *Journal of Geophysical Research*, **91**, 10557–10565.

Feller, W., 1968: *An Introduction to Probability Theory and its Applications*, Vol. 1, 3rd edn. New York, Wiley.

Fofonoff, N. P., 1954: Steady flow in a frictionless ocean. *Journal of Marine Research*, **13**, 254–262.

Foreman, M. G. G. and A. F. Bennett, 1988: On no-slip boundary conditions for the incompressible Navier-Stokes equation. *Dynamics of Atmospheres and Oceans*, **12**, 47–70.

Freeland, H. J., P. B. Rhines and T. Rossby, 1975: Statistical observations of the trajectories of neutrally buoyant floats in the North Atlantic. *Journal of Marine Research*, **33**, 383–404.

Friedman, A., 1964: *Partial Differential Equations of Parabolic Type*. Englewood Cliffs, Prentice-Hall.

Frisch, U., 1995: *Turbulence: The Legacy of A. N. Kolomogorov*. Cambridge, Cambridge University Press.

Fukumori, I., 2001: Data assimilation for models. In *Satellite Altimetry and Earth Sciences*, ed. L.-L. Fu and A. Cazanave. San Diego, Academic Press.

Gardner, C. W., 1985: *Handbook of Stochastic Methods*, 2nd edn. Berlin, Springer-Verlag.

Gargett, A. E., 1985: Evolution of scalar spectra with the decay of turbulence in a stratified fluid. *Journal of Fluid Mechanics*, **159**, 379–407.

Gargett, A. E., T. R. Osborn and P. Nasmyth, 1984: Local isotropy and the decay of turbulence in a stratified fluid. *Journal of Fluid Mechanics*, **144**, 231–280.

Garrett, C., J. Middleton, M. Hazen and F. Majaess, 1985: Tidal currents and eddy statistics from iceberg trajectories off Labrador. *Science*, **227**, 1333–1335.

Gelb, A. (ed.), 1974: *Applied Optimal Estimation*. Cambridge, MIT Press.

Gent, P. R. and J. C. McWilliams, 1990: Isopycnal mixing in ocean circulation models. *Journal of Physical Oceanography*, **20**, 150–155.

Ghil, M., S. Cohn, J. Tavantis, K. Bube and E. Isaacson, 1981: Applications of estimation theory to numerical weather prediction. In *Dynamical Meteorology, Data Assimilation Methods*, ed. L. Bengtsson, M. Ghil and E. Källén. Applied Mathematical Sciences No. 36, New York, Springer-Verlag.

Gibson, M. M., 1963: Spectra of turbulence in a round jet. *Journal of Fluid Mechanics*, **15**, 161–173.

Gilbarg, D. and N. S. Trudinger, 1983: *Elliptic Partial Differential Equations of Second Order*, 2nd edn. Berlin, Springer-Verlag.

Gill, A. E., 1982: *Atmosphere–Ocean Dynamics*. London, Academic Press.

Gnedenko, B., 1976: *The Theory of Probability*. Moscow, Mir.

Goldstein, S., 1965: *Modern Developments in Fluid Dynamics*, vols 1 and 2. New York, Dover.

Goodman, L. and E. R. Levine, 1990: Vertical motion of neutrally bouyant floats. *Journal of Atmospheric and Oceanic Technology*, **7**, 38–49.

Gould, W. J., 1998: Technology developments (Don't take them for granted). In *International WOCE Newsletter*, No. 30, WOCE International Planning Office, Southampton Oceanography Centre, Southampton.

Gould, W. J., 2001: Direct measurements of subsurface ocean currents: A success story. In *Understanding the Oceans: A Century of ocean exploration*, ed. M. Deacon, A. Rice and C. Summerhayes. London, University College of London Press.

Grant, H. L., R. W. Stewart and A. Moilliet, 1962: Turbulence spectra from a tidal channel. *Journal of Fluid Mechanics*, **12**, 241–263.

Guidry, M. W., 1991: *Gauge Field Theories: An Introduction with Applications*. New York, Wiley.

Guyon, E., J.-P. Hulin, L. Petit and C. D. Mitescu, 2001: *Physical Hydrodynamics*. Oxford, Oxford University Press.

Hadamard, J., 1923: *Lectures on Cauchy's Problem in Linear Partial Differential Equations*. New York, Yale University Press.

Haidvogel, D. B. and P. B. Rhines, 1983: Waves and circulation driven by oscillatory winds in an idealized ocean basin. *Geophysical and Astrophysical Fluid Dynamics*, **25**, 1–63.

Hill, R. J., 1978: Models of the scalar spectrum for turbulent advection. *Journal of Fluid Mechanics*, **88**, 541–562.

Holloway, G., 1994: Comment: on modeling vertical trajectories of phytoplankton in a mixed layer. *Deep-Sea Research*, **41**, 957–959.

Ide, K., L. Kuznetsov and C. K. R. T. Jones, 2002: Lagrangian data assimilation for point vortex systems. *Journal of Turbulence*, **3**, 1–6.

Il'in, A. M., A. S. Kalashnikov and O. A. Oleinik, 1962: Linear equations of the second order of parabolic type. *Russian Mathematical Surveys*, **17**, 1–143.

Ingber, L., 1993: Very fast simulate re-annealing. *Journal of Mathematical and Computational Modeling*, **12**, 967–973.

Jeffreys, H., 1931: *Cartesian Tensors*. Cambridge, Cambridge University Press.

Judovich, V. I., 1964: A two-dimensional problem of unsteady flow of an incompressible fluid across a given domain. *Matemacheskii Sbornik*, **64** (106), 562–588. (English translation: *Translations of the American Mathematical Society*, **57**, 277–304 (1966).)

Julian, P., W. Massman and N. Levanon, 1977: The TWERL experiment. *Bulletin of the American Meteorological Society*, **58**, 936–948.

Kamachi, M. and J. J. O'Brien, 1995: Continuous data assimilation of drifting buoy trajectory into an equatorial Pacific Ocean model. *Journal of Marine Systems*, **6**, 159–178.

Kamke, E., 1959: *Differentialgleichungen: Lösungsmethoden und Lösungen, Band 1: Gewöhnliche Differentialgleichungen*. New York, Chelsea Publishing Company.

Kato, T., 1967: On classical solutions of the two–dimensional non–stationary Euler equation. *Archive for Rational Mechanics and Analysis*, **25**, 188–200.

Kennan, S. C., P. P. Niiler and A. Sybrandys, 1998: Advances in drifting buoy technology. In *International WOCE Newsletter*, No. 30, WOCE International Planning Office, Southampton Oceanography Centre, Southampton.

Keppene, C. L., 2000: Data assimilation into a primitive-equation model with a parallel ensemble Kalman filter. *Monthly Weather Review*, **128**, 1971–1981.

Kinsman, B. 1965: *Wind Waves: Their Generation and Propagation on the Ocean Surface*. Englewood Cliffs, Prentice-Hall.

Kirwan, A. D., 1975: Oceanic velocity gradients. *Journal of Physical Oceanography*, **5**, 729–735.

Kirwan, A. D., 1988: Notes on the cluster method for interpreting relative motions. *Journal of Geophysical Research*, **93**, 9337–9339.

Klinker, E., F. Rabier, G. Kelly and J.-F. Mahfouf, 2000: The ECMWF operational implementation of four–dimensional variational assimilation. III: Experimental results and diagnostics with operational configurations. *Quarterly Journal of the Royal Meteorological Society*, **126**, 1191–1215.

Kraichnan, R. H., 1965: Lagrangian-history closure for turbulence. *Physics of Fluids*, **8**, 575–598.

Kraichnan, R. H., 1966a: Isotropic turbulence and inertial-range structure. *Physics of Fluids*, **9**, 1728–1752.

Kraichnan, R. H., 1966b: Dispersion of particle pairs in homogeneous turbulence. *Physics of Fluids*, **9**, 1937–1943.

Kraichnan, R. H., 1967: Inertial ranges in two-dimensional turbulence. *Physics of Fluids*, **10**, 1417–1423.

Kraichnan, R. H., 1968: Small-scale structure of a scalar field convected by turbulence. *Physics of Fluids*, **11**, 945–953.

Kraichnan, R. H., 1971: Inertial-range transfer in two- and three-dimensional turbulence. *Journal of Fluid Mechanics*, **47**, 525–535.

Kraichnan, R. H., 1974: Convection of a passive scalar by a quasi-uniform random straining field. *Journal of Fluid Mechanics*, **64**, 737–762.

LaCasce, J. H. and A. Bower, 2000: Relative dispersion in the subsurface North Atlantic. *Journal of Marine Research*, **58**, 863–894.

LaCasce, J. H. and C. Ohlmann, 2003: Relative dispersion at the surface of the Gulf of Mexico. *Journal of Marine Research*, **61**, 285–312.

Ladyzhenskaya, O. A., 1969: *The Mathematical Theory of Viscous Incompressible Flow*. Translated from the Russian by R. A. Silverman, 2nd English edition. New York, Gordon and Breach.

Lamb, H., 1932: *Hydrodynamics*, 6th edn. Cambridge, Cambridge University Press.

Lanczos, C., 1966: *The Variational Principles of Mechanics*, 3rd edn. Toronto, University of Toronto.

Landau, L. D. and E. M. Lifschitz, 1959: *Fluid Mechanics*. Translated from the Russian by J. B. Sykes and W. H. Reid. London, Pergamon Press.

Leblanc, S., 2004: Local stability of Gerstner's waves. *Journal of Fluid Mechanics*, **506**, 245–254.

Lee, J. H. W., and V. H. Chu, 2003: *Turbulent Jets and Plumes: a Lagrangian Approach*. Boston, Kluwer Academic Publishers.

Lesieur, M., and J. Herring, 1985: Diffusion of a passive scalar in two–dimensional turbulence. *Journal of Fluid Mechanics*, **161**, 77–95.

Lesieur, M., J. Sommeria and G. Holloway, 1981: Zones inertielles du spectre d'un contaminant passif en turbulence bidimensionelle. *Comptes Rendus des Seances d'Académie Scientifique, Série A*, **292–II**, 271–274.

Levins, R., 1969: The effect of random variation of different types of population growth. *Proceedings of the National Academy of Sciences of the U.S.A.*, **62**, 1061–1065.

Lien, R-C. and E. A. D'Asaro, 2002: The Kolmogorov constant for the Lagrangian velocity spectrum and structure function. *Physics of Fluids*, **14**, 4456–4459.

Lin, C. C., 1963: Hydrodynamics of Helium II. In *Liquid Helium*, Proceedings of the International Summer School of Physics "Enrico Fermi," Course XXI, Ed. G. Careri. New York, Academic Press.

Lin, J.-T., 1972: Relative dispersion in the enstrophy-cascading inertial range of homogeneous two-dimensional turbulence. *Journal of the Atmospheric Sciences*, **29**, 394–396.

Lumley, J. L., 1962: An approach to the Eulerian–Lagrangian problem. *Journal of Mathematical Physics*, **3**, 309–312.

Lumley, J. L. and H. A. Panofsky, 1964: *The Structure of Atmospheric Turbulence*. New York, Interscience Publishers.

Lundgren, T., 1981: Turbulent pair dispersion and scalar diffusion. *Journal of Fluid Mechanics*, **111**, 25–57.

Mahfouf, J.-F. and F. Rabier, 2000: The ECMWF operational implementation of four-dimensional variational assimilation. II: Experimental results with improved physics. *Quarterly Journal of the Royal Meteorological Society*, **126**, 1171–1190.

Majda, A. J. and A. L. Bertozzi, 2002: *Vorticity and Incompressible Flow*. Cambridge, Cambridge University Press.

Malanotte–Rizzoli, P. (ed.), 1996: *Modern Approaches to Data Assimilation in Ocean Modeling*. Amsterdam, Elsevier.

McGrath, F. J., 1968: Nonstationary plane flow of viscous and ideal fluids. *Archive for Rational Mechanics and Analysis*, **27**, 329–348.

McPhaden, M. J., 1995: The Tropical–Atmosphere Ocean array is completed. *Bulletin of the American Meteorological Society*, **76**, 739–742.

McWilliams, J. C. et al., 1983: The local dynamics of eddies in the western North Atlantic. In *Eddies in marine Science*, ed. A. R. Robinson. Berlin, Springer-Verlag.

Mead, J. L., 2004: The shallow-water equations in Lagrangian coordinates. *Journal of Computational Physics*, **200**, 654–669.

Mead, J. L., 2005: Assimilation of simulated float data in Lagrangian coordinates. *Ocean Modelling*, **8**, 369–394.

Mead, J. L. and A. F. Bennett, 2001: Towards regional assimilation of Lagrangian data: The Lagrangian form of the shallow water reduced gravity model and its inverse. *Journal of Marine Systems*, **29**, 365–384.

Middleton, J. F., 1980a: A review of relative diffusion analysis and results. *Journal of Physical Oceanography*, **10**, 992–996.

Middleton, J. F., 1980b: Reply to Comments by A. D. Kirwan, Jr., and W. J. Merrell, Jr., on "A review of relative diffusion analysis and results". *Journal of Physical Oceanography*, **10**, 995–996.

Middleton, J. F., 1985: Drifter spectra and diffusivities, *Journal of Marine Research*, **43**, 37–55.

Middleton, J. F. and C. Garrett, 1986: A kinematic analysis of polarized eddy fields using drifter data. *Journal of Geophysical Research*, **91**, 5094–5102.

Middleton, J. F. and J. W. Loder, 1989: Skew fluxes in polarized wave fields. *Journal of Physical Oceanography*, **19**, 68–76.

Miller, R. N., 1986: Toward the application of the Kalman filter to regional ocean modeling. *Journal of Physical Oceanography*, **16**, 72–86.

Milne-Thomson, L. M., 1960: *Theoretical Hydrodynamics*. New York, Macmillan.

Molinari, R., A. D. Kirwan, 1975: Calculations of differential kinematic properties from Lagrangian observations in the Western Caribbean Sea. *Journal of Physical Oceanography*, **5**, 483–491.

Monin, A. S., 1962: Lagrangian equations in the hydrodynamics of an incompressible viscous fluid. *Prikladaniya Matemacheskii i Mechanika*, **26**, 320–327.

Monin, A. S. and A. M. Yaglom, 1971: *Statistical Fluid Mechanics: Mechanics of Turbulence*, Vol. I, ed. J. Lumley. Cambridge, MIT Press.

Monin, A. S. and A. M. Yaglom, 1975: *Statistical Fluid Mechanics: Mechanics of Turbulence*, Vol. II, ed. J. Lumley. Cambridge, MIT Press.

Morel, P. and M. Larcheveque, 1974: Relative dispersion of constant-level balloons in the 200mb general circulation. *Journal of the Atmospheric Sciences*, **31**, 2189–2196.

Mory, M. and E. J. Hopfinger, 1986: Structure functions in a rotationally dominated turbulent flow. *Physics of Fluids*, **29**, 2140–2146.

Nakahara, M., 1990: *Geometry, Topology and Physics*. Bristol, Adam Hilger.

Niiler, P. P., 2001: The world ocean surface circulation. In *Ocean Circulation and Climate: Observing and Modelling the Global Ocean*, ed. G. Siedler, J. Church and J. Gould. San Diego, Academic Press.

Niiler, P. P., R. E. Davis and H. J. White, 1987: Water-following characteristics of a mixed layer drifter. *Deep-Sea Research*, **34**, 1867–1881.

Niiler, P. P., A. L. Sybrandy, K. Bi, P. Poulain and D. Bitterman, 1995: Measurements of the water-following capabilities of Holey-sock and TRISTAR drifters. *Deep-Sea Research*, **42**, 1951–1964.

Obhukov, A. M., 1949: Structure of the temperature field in a turbulent flow. *Izvestiia Akademii Nauk SSSR. Seriia Geograficheskaia i Geofizicheskaia.* **13**, 58–59.

Okubo, A., 1971: Oceanic diffusion diagrams. *Deep-Sea Research*, **18**, 789–802.

Okubo, A. and C. C. Ebbesmeyer, 1975: Determination of vorticity, divergence and deformation rates from analysis of drogue observations. *Deep-Sea Research*, **23**, 349–352.

Okubo, A., C. C. Ebbesmeyer and J. M. Helseth, 1976: Determination of Lagrangian deformations from analysis of current followers. *Journal of Physical Oceanography*, **6**, 524–527.

Oliger, J. and A. Sundström, 1978: Theoretical and practical aspects of some initial boundary value problems in fluid dynamics. *SIAM Journal of Applied Mathematics*, **35**, 4119–446.

Ollitrault, M., 1993: The SAMBA0 experiment: a comparison of subsurface floats using the RAFOS technique. *WOCE Newsletter*, **14**, 28–32 (unpublished manuscript).

Ottino, J. M., 1989: *The Kinematics of Mixing: Stretching, Chaos and Transport.* Cambridge, Cambridge University Press.

Pais, A., 1982: *'Subtle is the Lord...' The Science and the Life of Albert Einstein.* Oxford, Oxford University Press.

Pedlosky, J., 1987: *Geophysical Fluid Dynamics*, 2nd edn. New York, Springer-Verlag.

Perry, A. E. and M. S. Chong, 1987: A description of eddying motions and flow patterns using critical-point concepts. *Annual Reviews of Fluid Mechanics*, **19**, 125–155.

Phillips, O. M., 1966: *The Dynamics of the Upper Ocean.* Monographs on Applied Mathematics and Mechanics, Cambridge, Cambridge University Press.

Pollard, R. T., 1970: Surface waves with rotation: an exact solution. *Journal of Geophysical Research*, **75**, 5895–5898.

Pond, S., R. W. Stewart and R. W. Burling, 1963: Turbulence spectra in the wind over waves. *Journal of the Atmospheric Sciences*, **20**, 319–324.

Pope, S. B., 1994: Lagrangian pdf methods for turbulent flows. *Annual Reviews of Fluid Mechanics*, **26**, 23–63.

Prigogine, I., 1980: *From Being to Becoming: Time and Complexity in the Physical Sciences.* San Francisco, W. H. Freeman.

Rabier, F., H. Järvinen, E. Klinker, J.-H. Mahfouf and A. Simmons, 2000: The ECMWF operational implementation of four-dimensional assimilation. I: Experimental results with simplified physics. *Quarterly Journal of the Royal Meteorological Society*, **126**, 1143–1170.

Rankine, W. J. M., 1863: On the exact form of waves near the surface of deep water. *Philosophical Transactions of the Royal Society of London*, **153**, 127–138.

Rhines, P. B., 1977: The dynamics of unsteady currents. In *The Sea*, Vol. 6: Marine Modelling, ed. E. D. Goldberg, I. N. McCave, J. J. O'Brien and J. H. Steele. New York, Wiley.

Richardson, L. F., 1926: Atmospheric diffusion shown on a distance-neighbour graph. *Proceedings of the Royal Society of London, Series A*, **110**, 709–737.

Ripa, P., 1981: Symmetries and conservation laws for internal gravity waves. *American Institute of Physics Proceedings*, **76**, 281–306.

Rodean, H. C., 1996: *Stochastic Lagrangian models of turbulent diffusion*. Meteorological Monographs, **26**, No. 48, American Meteorological Society, Boston.

Rosenhead, L., 1963: *Laminar Boundary Layers*. Oxford, Clarendon Press.

Rossby, T., D. Dorson and J. Fontaine, 1986: The RAFOS system. *Journal of Atmospheric and Oceanic Technology*, **3**, 672–679.

Rossby, H. T., E. R. Levine and D. N. Connors, 1985: The isopycnal Swallow float – a simple device for tracking water parcels in the ocean. In *Essays on Oceanography: a tribute to John Swallow*, ed. J. Crease, W. J. Gould and P. M. Saunders, *Progress in Oceanography*, **14**, 511–525.

Rossby, T., A. D. Voorhis and D. Webb, 1975: A quasi-Lagrangian study of mid-ocean variability using long range SOFAR floats. *Journal of Marine Research*, **33**, 355–382.

Ryder, L. H., 1996: *Quantum Field Theory*, 2nd edn. Cambridge, Cambridge University Press.

Salmon, R., 1983: Practical use of Hamilton's principle. *Journal of Fluid Mechanics*, **132**, 431–444.

Samelson, R. M., and S. Wiggins, 2005: *Lagrangian Transport in Geophysical Jets and Waves*. New York, Springer-Verlag.

Sanderson, B. G., B. K. Pal and A. Goulding, 1988: Calculations of unbiased estimates of the magnitude of residual velocities from a small number of drogue trajectories. *Journal of Geophysical Research*, **93**, 8161–8162.

Sattinger, D. H. and O. L. Weaver, 1986: *Lie Groups and Algebras with Applications to Physics, Geometry, and Mechanics*. New York, Springer-Verlag.

Saucier, W. J., 1955: *Principles of Meteorological Analysis*. Chicago, University of Chicago Press.

Schlichting, H., 1960: *Boundary Layer Theory*. New York, McGraw-Hill.

Schulman, L. S., 1981: *Techniques and Applications of Path Integration*. New York, John Wiley.

Seliger, R. L. and G. B. Witham, 1968: Variational principles in continuum mechanics. *Proceedings of the Royal Society of London*, Series A, **305**, 1–25.

Sreenivasan, K. R., 1995: On the universality of the Kolmogorov constant. *Physics of Fluids*, **7**, 2778–2784.

Stammer, D., C. Wunsch, R. Giering, C. Eckart, P. Heimbach. J. Marotzke, A. Adcroft, C. N. Hill and J. Marshall, 2003: Volume, heat and freshwater transports of the global ocean circulation 1993–2000, estimated from a general circulation model constrained by World Ocean Circulation Experiment (WOCE) data. *Journal of Geophysical Research*, **108**, 3007–3028.

Stoker, J. J., 1957: *Water Waves*. New York, Interscience Publishers.

Stuart, J. T., 1987: Nonlinear Euler partial differential equations: Singularities in their solution. In *Symposium to Honor C. C. Lin*, edited D. J. Benney, F. H. Shu and C. Yuan, 81–95. Singapore, World Scientific.

Stuart, J. T., 1998: Singularities in three-dimensional compressible Euler flows. *Theoretical and Computational Fluid Dynamics*, **10**, 385–391.

Sullivan, P. J., 1971: Some data on the distance–neighbour function for relative diffusion. *Journal of Fluid Mechanics*, **47**, 601–607.

Sverdrup, H. U., M. W. Johnson and R. H. Fleming, 1942: *The Oceans: Their Physics, Chemistry and General Biology*. Englewood Cliffs, Prentice-Hall.

Swallow, J. C., 1955: A neutral-buoyancy float for measuring deep currents. *Deep-Sea Research*, **3**, 74–81.

Swallow, J. C., 1969: A deep current of Cape St Vincent. *Deep-Sea Research*, **16** (Supplement), 285–285.

Swallow, J. C., and L. V. Worthington, 1961: An observation of a deep countercurrent in the Western North Atlantic. *Deep-Sea Research, 8*, 1–19.

Talagrand, O. and P. Courtier, 1987: Variational assimilation of meteorological observations with the adjoint vorticity equation. I: Theory. *Quarterly Journal of the Royal Meteorological Society*, **113**, 1311–1328.

Taylor, G. I., 1921: Diffusion by continuous movements. *Proceedings of the London Mathematical Society, Series 2*, **20**, 196–211.

Temam, R., 2000: Some developments on Navier–Stokes equations in the second half of the 20th century. In *Development of Mathematics 1950–2000*, ed. J.-P. Pier. Berlin, Birkhaüser.

Tennekes, H. and J. L. Lumley, 1972: *A First Course in Turbulence*. Cambridge, MIT Press.

Thompson, C., 1988: *Classical Equilibrium Statistical Mechanics*. New York, Oxford University Press.

Tizard, T. H., H. N. Moseley, J. Y. Buchanan and J. Murray, 1985: Narrative of the cruise of HMS *Challenger*, with a general account of the scientific results of the expedition. In *Report on the scientific results of the voyage of HMS Challenger during the years 1872–76* (1880–95), **1**, ed. C. W. Thompson and J. Murray. Her Majesty's Stationery Office, London.

Tokaty, G. A., 1971: *A History and Philosophy of Fluidmechanics*. Henley–on–Thames, Foulis.

Truesdell, C., 1954: *The Kinematics of Vorticity*. Bloomington, Indiana University Press.

van Kampen, N. G., 1992: *Stochastic Processes in Physics and Chemistry*. Amsterdam, North–Holland.

Weinberg, S., 1995: *The Quantum Theory of Fields, II: Modern Applications*. Cambridge, Cambridge University Press.

Whittaker, E. T. and G. N. Watson, 1927: *Modern Analysis*, 4th edn. Cambridge, Cambridge University Press.

Williams, R. M. and C. A. Paulson, 1977: Microscale temperature and velocity spectra in the atmospheric boundary layer. *Journal of Fluid Mechanics*, **83**, 547–567.

WMO, 1977: *A Summary of the FGGE Implementation Plan*. World Meteorological Organization, Geneva.

Wolibner, W., 1933: Un theorème sur l'existence du mouvement plan d'un fluide parfait, homogène, incompressible, pendant un temps infiniment long. *Mathematische Zeitschrift*, **37**, 698–726.

Wunsch, C., 1996: *The Ocean Circulation Inverse Problem*. New York, Cambridge University Press.

Yakubovich, E. I. and D. A. Zenkovich, 2001: Matrix approach to Lagrangian fluid dynamics. *Journal of Fluid Mechanics*, **443**, 167–196.

Yakubovich, E. I. and D. A. Zenkovich, 2002: Matrix fluid dynamics. In *International Conference on Progress in Nonlinear Science*, Vol.2, ed. A. G. Litvak and N. Novgorod, pp 282–287. e–Print arXiv:physics/00110004.

Yamazaki, H. and D. Kamykowski, 1991: The vertical trajectories of motile phytoplankton in a wind-mixed water column. *Deep-Sea Research*, **32**, 219–241.

Yamazaki, H. and D. Kamykowski, 1994: Reply to Greg Holloway. *Deep-Sea Research*, **41**, 961–963.

Young, W. R., P. B. Rhines and C. J. R. Garrett, 1982: Shear-flow dispersion, internal waves and horizontal mixing in the ocean. *Journal of Physical Oceanography*, **12**, 515–527.

Zakharov, V. E. and E. A. Kuznetsov, 1997: Hamiltonian formalism for nonlinear waves. *Physics-Uspekhi*, **40**, 1087–1116.

Subject index

Author index

For EU product safety concerns, contact us at Calle de José Abascal, 56–1°,
28003 Madrid, Spain or eugpsr@cambridge.org.

www.ingramcontent.com/pod-product-compliance
Lightning Source LLC
LaVergne TN
LVHW011525200926
848484LV00001B/159
* 9 7 8 0 5 2 1 8 5 3 1 0 1 *